# Blue Space, Health and Wellbeing

T0187438

Health geography makes critical contributions to contemporary and emerging interdisciplinary agendas of nature-based health and health-enabling places. Couched in theory and critical empirical work on nature and health, this book addresses questions on the relationships between water, health and wellbeing. Water and blue space is a key focus in current health geography research and a new *hydrophilic turn* has emerged with a particular focus on the aspects of water which are affective, life-enhancing and health-enabling. Research considers the benefits and risks associated with blue space, from access to safe and clean water in the Global South, to health promoting spaces found around urban waters, to the deeper implications of climate change for water-based livelihoods and indigenous cultures. This book reflects recent theoretical debates within health geography, drawing from research in the public health, anthropology and psychology sectors. Broad thematic sections focus on interdisciplinary, experiential and equity-based elements of blue space, with individual chapters that consider indigenous and global health, water's healing properties, leisure and blue yogic culture, coastal landscapes, surfing, swimming and sailing, along with more contested hydrophobic dimensions.

The interdisciplinary lens means this book will be extremely valuable to human geographers and cultural geographers. It will also appeal to practitioners and researchers interested in environmental health, leisure and tourism, health inequalities and public health more broadly.

**Ronan Foley** is Senior Lecturer in the Department of Geography at Maynooth University, Ireland.

**Robin Kearns** is Professor of Geography in the School of Environment at the University of Auckland, New Zealand.

**Thomas Kistemann** is Professor of Hygiene, Environmental Medicine and Medical Geography at the University of Bonn, Germany.

**Ben Wheeler** is Senior Lecturer at the University of Exeter Medical School, UK.

# Geographies of Health

Edited by **Allison Williams,** *Associate Professor, School of Geography and Earth Sciences, McMaster University, Canada, and* **Susan Elliott,** *Professor, Department of Geography and Environmental Management and School of Public Health and Health Systems, University of Waterloo, Canada*

There is growing interest in the geographies of health and a continued interest in what has more traditionally been labeled medical geography. The traditional focus of 'medical geography' on areas such as disease ecology, health service provision and disease mapping (all of which continue to reflect a mainly quantitative approach to inquiry) has evolved to a focus on a broader, theoretically informed epistemology of health geographies in an expanded international reach. As a result, we now find this subdiscipline characterized by a strongly theoretically-informed research agenda, embracing a range of methods (quantitative; qualitative and the integration of the two) of inquiry concerned with questions of: risk; representation and meaning; inequality and power; culture and difference, among others. Health mapping and modeling has simultaneously been strengthened by the technical advances made in multilevel modeling, advanced spatial analytic methods and GIS, while further engaging in questions related to health inequalities, population health and environmental degradation.

This series publishes superior quality research monographs and edited collections representing contemporary applications in the field; this encompasses original research as well as advances in methods, techniques and theories. The *Geographies of Health* series will capture the interest of a broad body of scholars, within the social sciences, the health sciences and beyond.

**Public Health, Disease and Development in Africa**
*Edited by Ezekiel Kalipeni, Juliet Iwelunmor, Diana Grigsby-Toussaint and Imelda K. Moise*

**Blue Space, Health and Wellbeing**
Hydrophilia Unbounded
*Edited by Ronan Foley, Robin Kearns, Thomas Kistemann and Ben Wheeler*

For a full list of titles in this series, please visit https://www.routledge.com/ Geographies-of-Health-Series/book-series/GHS

# Blue Space, Health and Wellbeing

## Hydrophilia Unbounded

**Edited by Ronan Foley, Robin Kearns, Thomas Kistemann and Ben Wheeler**

Routledge
Taylor & Francis Group

LONDON AND NEW YORK

First published 2019
by Routledge
2 Park Square, Milton Park, Abingdon, Oxon OX14 4RN

and by Routledge
52 Vanderbilt Avenue, New York, NY 10017

First issued in paperback 2020

*Routledge is an imprint of the Taylor & Francis Group, an informa business*

*British Library Cataloguing-in-Publication Data*
A catalogue record for this book is available from the British Library

*Library of Congress Cataloging-in-Publication Data*
A catalog record has been requested for this book

ISBN 13: 978-0-367-66180-9 (pbk)
ISBN 13: 978-0-8153-5914-2 (hbk)

Typeset in Times New Roman
by codeMantra

We collectively dedicate this book to Wilbert Gesler who, through his articulation of a cultural geography of health and development of the therapeutic landscape idea, opened wide the horizons of our sub-discipline.

# Contents

# Figures

# Tables

# Boxes

# Notes on contributors

**Carmen Anthonj** is a Postdoctoral Research Associate at The Water Institute at the Gillings School of Global Public Health at the University of North Carolina, Chapel Hill, USA. Her background is in Medical Geography, and her research interests cover the interlinkages between the environment, water and human health with a major focus on water-related infectious diseases, water, sanitation and hygiene (WASH); health risk perceptions and health-related behaviour; and the cultural context of health. Carmen enjoys applying mixed methods and multidisciplinary perspectives. She holds a doctoral degree from the GeoHealth Centre at Institute for Hygiene and Public Health at the University of Bonn, Germany, and conducted research with the World Health Organization and UNICEF, and several governments and research institutes. Carmen has health risk- and WASH-related field experience in Sub-Saharan Africa, South America and Southeast Asia, and in the South Pacific.

**Sarah Atkinson** has a background in Anthropology and Human Nutrition and is Professor of Geography and Medical Humanities at Durham University, UK. Her research interrogates the hidden assumptions, social meanings and implications embedded in key contemporary health concepts such as wellbeing and care. She is part of the Economic and Social Research Council's What Works Centre for Wellbeing evidence programme on communities and a member of the WHO Europe Expert Advisory Group on the Cultural Contexts of Health. Her publications include *The Edinburgh Companion to the Critical Medical Humanities* (Whitehead, Wood, Atkinson, Macnaughton and Richards, eds.) and *Wellbeing and Place* (Atkinson, Fuller and Painter, eds.).

**Sarah L. Bell** is Research Fellow at the University of Exeter, UK. Her research focuses on the complex intersections between human health, wellbeing and the interlinked physical, social and cultural environments encountered through the life-course, including those often characterised as 'blue' or 'green'. Sarah's work is underpinned by a passion for qualitative methodological development, designing sensitive approaches that promote critical awareness of alternative ways of embodying, experiencing and interpreting diverse everyday geographies.

**Easkey Britton** is a postdoctoral research fellow at the Whitaker Institute, NUI Galway, Ireland, where she specialises in research on oceans and human health. She is best known as a big-wave surfer from Ireland, with a PhD in environment and society, who is channelling her passion for surfing into social change. Britton convenes and facilitates workshops on leadership and social impact globally. Presenter and co-producer of the award winning documentary on women surfing in Iran, 'Into the Sea,' Britton recently founded Like Water, a social change initiative that seeks to foster positive relationships with water and the sea.

**Mike Brown** has published several pieces on the sea and human experiences. He has co-edited *Seascapes: Shaped by the Sea* with Barbara Humberstone (2015) and *Living with the Sea* with Kimberley Peters (2018). He is currently General Manager for Coastguard Boating Education in Auckland, New Zealand, and lives aboard his boat in an inner-city marina.

**Michael H. Depledge** (CBE, DSc, FRCP) is Chair of Environment and Human Health at the University of Exeter Medical School, UK. He was trained as a marine ecotoxicologist, before moving into cancer research. He has conducted studies in oceans and human health for over 35 years, particularly in Southeast Asia, Scandinavia, and the Americas. His main interests include using the marine environment to foster improvements in health and wellbeing (including the development of the 'Blue Gym' approach), as well as climate change and emerging chemical threats in the sea. Michael was previously the Chief Scientist of the UK Environment Agency, and is an adviser to the Intergovernmental Oceanographic Commission, International Maritime Organisation, United Nations Environment Programme and World Health Organisation.

**Karolina Doughty** is a lecturer in cultural geography at Wageningen University and Research, the Netherlands. She is a cultural geographer with interests that span cultural, urban, health and tourism geographies. Karolina's research broadly explores the microgeographies of wellbeing, focusing on how wellbeing is manifest in everyday places and practices, particularly those processes that render certain places as 'therapeutic'. Her research is often framed by an attention to embodied movement, emotion and affect and sensory experience, through creative, mobile and visual methods.

**Timo Falkenberg** is the coordinator of the Forschungskolleg 'One Health and Urban Transformation' and a Lecturer in Geography as well as Social Medicine at the University of Bonn, Germany. He is a Public Health researcher and a Medical Geographer based at the Zentrum für Entwicklungsforschung/Center for Development Research (ZEF) with a special interest in waterborne disease transmission. During his doctoral research he focused on urban agriculture, irrigation and health in Ahmedabad, India.

**Ronan Foley** is Senior Lecturer in the Department of Geography at Maynooth University, Ireland. He is a health geographer with specific interest in healthcare services, caring, mental health and therapeutic landscapes. His most recent work explores experiential, embodied and emotional dimensions of 'healthy blue space'. He co-edited, with Thomas Kistemann, a 2015 special issue on healthy blue space, for the journal *Health & Place* and has also authored *Healing Waters: Therapeutic Landscapes in Historic and Contemporary Ireland* (2010) as part of the Ashgate/Routledge Geographies of Health book series. He is the current editor of *Irish Geography* and was an Erskine Fellow in the Department of Geography at the University of Canterbury in Spring 2015.

**Sean Gammon** is based in the School of Management at the University of Central Lancashire, UK. He is widely published in sport-related tourism, primarily focusing on customer motivation, nostalgia and heritage. In addition, he continues to contribute to the field of leisure, recently co-editing a new text on *Leisure Landscapes*, published by Routledge.

**Julie Hollenbeck** is an applied qualitative researcher at the University of Exeter, UK, with expertise in environment and human health and scientific outreach. She is primarily interested in reducing health disparities, having recently completed a PhD exploring recreational marine-based ecosystem service access inequities in an African-American community in Miami. Dr Hollenbeck approaches her work using a multifaceted lens, taking into account individual/group relationships to the environment, through cultural, social, historical, political and racial/ethnic contexts, and the impact this has on human health and wellbeing.

**David Jarratt** is a Senior Lecturer in the School of Management at the University of Central Lancashire, UK. His research interests revolve around sense of place and the consumption of tourism and leisure experiences, especially those relating to wellness. In recent years, he has been focusing on coastal tourism but also has a long-standing interest in heritage tourism.

**Robin Kearns** is a Professor of Geography in the School of Environment at the University of Auckland, New Zealand. His research into the links between place and health was consolidated in the books *Putting Health into Place: Landscape, Identity and Wellbeing* (1998) and *Culture, Place and Health* (2002) (both with Wilbert Gesler). Recently, his work has taken a 'blue turn' with a focus on the significance of coastal settings to human wellbeing. In particular, in 2016 he edited a special issue of the *New Zealand Geographer* on islands and is currently leading an investigation into the wellbeing of older residents of Auckland's largest island. His previous books in the Ashgate/Routledge Geographies of Health series include *Soundscapes of Wellbeing in Popular Music* (2014, with G. Andrews and P. Kingsbury) and *Afterlives of the Psychiatric Asylum* (2015, with G. Moon and A. Joseph).

**Thomas Kistemann** has a background in medicine, geography and classic philology. He is a Professor of Hygiene, Environmental Medicine and Medical Geography at the University of Bonn, Germany, and is the head of the GeoHealth Centre at the Institute for Hygiene and Public Health as well as of the WHO Collaborating Centre for Health Promoting Water Management and Risk Communication. His extensive research interests include environmental health with a focus on blue environments, health landscapes, infectious disease epidemiology and public health applications of GIS. At WHO Europe, he is a member of the Working Group on Water & Health, and of the Expert Advisory Group on the Cultural Contexts of Health. He co-edited a special issue in 2015 on Healthy Blue Space (with Ronan Foley) for the journal *Health & Place* and a monograph in 2016 (with Ulrich Gebhard) on landscape, identity and health (in German).

**Rebecca Lovell** is a socio-environmental research fellow at the University of Exeter Medical School's European Centre for Environment and Human Health, UK. Her work considers the linkages between natural environments and health, wellbeing and quality of life outcomes and how these may be modified by sociocultural contexts or previous experiences. She is currently focusing on the practice of settings-based health delivery and 'nature'-based health interventions. She works with government departments and agencies to translate evidence to inform policy development and service delivery, most recently in relation to the benefits and threats of coastal living to health.

**Antony Lyons** operates as an independent creative practitioner and water permeates all of his landscape-based projects, while water issues have loomed large throughout his background of diverse transdisciplinary activities. These have transitioned through geology, environmental studies, landscape design, public-art sculpture, to site-specific intermedia installation. Today, much of his work engages deeply with the affective and aesthetic experience of water settings, including eco-metaphorical, ritual and sacred connections. He seeks knowledges that lie largely inaccessible and unapproachable, unless approached through lenses of intuition, imagination and poetics. As well as developing his own site-inspired projects, he uses the medium of film to highlight and communicate other inspiring, transformational water-based initiatives.

**Morgan Parnell,** SSW, currently pursuing a Bachelor of Social Work at McMaster University, Canada. While completing her BSW, Morgan works as the program coordinator at Common Compass, a non-profit organization that delivers evidence-based programs and workshops to high school students around social-emotional learning. She has a diverse background with volunteer and paid experience in youth justice and youth homelessness, research and policy analysis, and mental health and wellness.

**Meg Parsons** is a Senior Lecturer in geography at the School of Environment, University of Auckland, New Zealand. Her research is interdisciplinary in scope and nature, and crosses the boundaries between human geography, historical studies and Indigenous studies. That research also examines how different values and belief systems are translated into environmental policies and actions, the ways in which specific historical narratives and discourses influence both the construction and practices of scientific knowledge, and how colonialism influences contemporary Indigenous societies' responses to climate change and other environmental changes. She has published on environmental histories and on indigenous responses to climate change and was a co-author of a key National Climate Change Adaptation Research Plan for Indigenous Communities in Australia in 2013.

**Ashleigh Patterson** is a graduate student and teaching assistant at McMaster University, Ontario, Canada, completing her research on the development of the new Carer-Inclusive and Accommodating Organizations Standard with the Canadian Standards Association. She is the recipient of a Canadian SSHRC funding award to support that study and won two undergraduate research awards while studying for a degree in geography and earth science at McMaster University.

**Katherine Phillips** is a researcher at the University of the West of England, UK. She is particularly interested in human-environment relationships. In her role in the large Arts and Humanities Research Council funded project *Towards Hydrocitizenship*, she has explored these relationships through social and community interactions and interventions around hidden rivers and the daylighting of them, human relationships with watery species such as eels, the hidden infrastructures of water and creative ways of engaging with the tides and tidal landscapes.

**Hannah Pitt** is a researcher at the Sustainable Places Research Institute, Cardiff University, UK, which specialises in interdisciplinary sustainability science. She is a social and cultural geographer, focusing on everyday experiences of blue and green spaces, and interactions with nonhumans. Previous research has taken a critical perspective on community gardens as places potentially enhancing wellbeing and promoting care ethics. Much of her research is delivered in collaboration with voluntary and community associations in the UK, including evaluations of the sustainability potential of third-sector food initiatives. Hannah specialises in innovative methodologies which enable both people and nonhumans to fully participate in social research, publishing on this topic with a focus on plants.

**Veronica Strang** is an environmental anthropologist and directs Durham University's Institute of Advanced Study, UK. Her research focuses on human-environment relations, especially societies' engagements with water. She has held academic positions at the University of Oxford, the University of Wales, Goldsmiths University and the University of Auckland.

From 2013 to 2017, she chaired the Association of Social Anthropologists of the UK and the Commonwealth. In 2000, she received a Royal Anthropological Institute Urgent Anthropology Fellowship, and in 2007 an international water prize from UNESCO. Her key publications include *Uncommon Ground: Cultural Landscapes and Environmental Values* (1997), *The Meaning of Water* (2004), *Gardening the World: Agency, Identity and the Ownership of Water* (2009), *Ownership and Appropriation* (2010) and *Water: Nature and Culture* (2015). She is currently writing a major comparative text examining historical and contemporary beliefs about water deities.

**Ben Wheeler** is a Senior Lecturer at the University of Exeter Medical Schools, UK. His main research interest is the interplay between environment, socio-economic status and public health, with particular regard to health inequalities and informing health and environmental policy. Key methodologies include the application of geographic information systems and spatial analysis to study these issues and the use of large, national and international secondary data sets. He is also interested in methods that apply global positioning system (GPS) and other technologies in the study of environment and health and wellbeing. Dr Wheeler works as part of a wider multidisciplinary group carrying out research on nature, health and wellbeing, investigating the potentially positive effects of natural environments on population health. He also works with collaborators in the European Centre, across the University of Exeter, and at other institutions to apply geographical methods to the investigation of a wide variety of environmental impacts on health.

**Mat White** is a social and environmental psychologist and Senior Lecturer, based at the University of Exeter Medical School, UK. He is primarily interested in how time use affects people's psychological wellbeing. Following evidence that time spent in natural, especially aquatic, environments is associated with particularly high levels of positive affect, he has spent the last few years trying to better understand why this might be.

**Allison Williams** is currently an Associate Professor in the School of Geography and Earth Sciences at McMaster University, Ontario, Canada. Trained as a health geographer, Allison works in many interdisciplinary groups, examining issues related to health, quality of life and wellbeing. In addition to keeping a pulse on applications in therapeutic landscape research, she currently holds a Canadian Institutes for Health Research (CIHR) Research Chair in Gender, Work and Health, and is examining how workplaces can best accommodate employees who are also juggling the role of family caregiving for adult children and the elderly. She currently supervises a team of seven trainees and works collaboratively with a wide range of partner organisations.

# Acknowledgements

The editors would like to thank a number of different agencies, groups, organisations and individuals for their help in putting this book together. First, we are very grateful to the Geographies of Health Series Editors, Susan Elliott and Allison Williams, for their encouragement of the submission of the initial idea and for excellent critical advice which allowed us to sharpen up that submission. We are in turn very grateful to Routledge, Taylor and Francis for running with the idea and for helping us pull it together, especially to Faye Leerink in the initial stages and then to Ruth Anderson and Emeline Jarvie for their sterling work on the final version and the production of the book. We are also very grateful to all the individual authors who contributed to the book and took time out of very busy and very different lives, to provide the interdisciplinary contribution we were striving for. The narratives came from many corners of the world and from many different physical and philosophical positions, but all shared a passion for water and its manifold ways of improving health and wellbeing.

We would also like to thank funding agencies whose support enabled us to take the time to develop the ideas contained here. For Ronan this includes the Environmental Protection Agency (Ireland), while Ben acknowledges the support of the BlueHealth project funded from the European Union's Horizon 2020 research and innovation programme under grant agreement no. 666773. In addition, we would very much like to record our appreciation of our home institutions, departments and colleagues for providing the background time and support to allow us to complete the book. These include the Department of Geography at Maynooth University and the Maynooth University Social Sciences Institute (MUSSI), the School of Environment at the University of Auckland, the Geography Department, Institute for Hygiene and Public Health at the University of Bonn and the European Centre for Environment and Human Health at the University of Exeter Medical School.

On a personal level, we would like to especially thank friends, family, colleagues and partners for their support. For Ronan this includes supportive colleagues in Maynooth and from the NEARHealth and ECOHealth projects in NUI Galway and UCD who share an interest in green and especially blue space. Ronan dedicates the book to Nell, whose own beautiful words more poetically describe the sea than he could ever think of. Robin would like to acknowledge conversations with Mike Brown, Kimberley Peters, Gregory O'Brien, Tara Coleman and Erin James, all of which have deepened a sense of the blue, as have regular commutes to Waiheke Island. For Thomas, this includes the distinguished colleagues and students from many different disciplines who contribute to the productive, vibrant and inspiring atmosphere at the IHPH GeoHealth Centre. Ben would like to thank all the excellent colleagues, students and collaborators of the European Centre for Environment and Human Health who share a passion for developing robust and useful evidence on interconnections between our health and natural environments. Ben dedicates this book to Jo Polack, whose creativity in response to the sea and our relationship with it is a daily inspiration.

# 1  Introduction

*Ronan Foley, Robin Kearns,*
*Thomas Kistemann and Ben Wheeler*

## Part A: hydrophilic beginnings

The time is right for a new text on relationships between water, health and wellbeing. Across many different subjects, there is an enhanced orientation towards the water, and a new *hydrophilic turn* has emerged with a particular focus on aspects of water that are affective, life-enhancing and health-enabling. Within the wider discipline, recent cultural geographies have reflected increasing public interest in writing around all things blue, including oceans and the sea, water-based sports and leisure geographies (Anderson and Peters, 2014); but also wider water-focused environments as sites of economic, imaginative and theoretical concern, especially in relation to climate change (Brown and Humberstone, 2015; Ryan, 2012). These authors speak to a range of perspectives oriented towards, by, from, on and in the water, that attest to its importance as a relational space and how it emerges as a polyvocal subject for ongoing discussions around nature and culture, the human and more-than-human (Abram, 1996; Gesler, 1992).

While more implicit than explicit, these wider cultural geographies have drawn attention to health in various guises, with a more explicit focus evident in recent texts on environmental health, technology and planning (Bartram, 2017). What makes the present book different is its focus on the relationships between water, health and wellbeing with an innate assumption of a hydrophilic leaning as a starting point. This perspective is developed by a range of researchers who fall loosely, though not fully, within the ambit of health geographies. Recent research within the subfield has been developed through therapeutic landscape studies. These studies have built on the foundations laid down by Wilbert Gesler (1992) and have served as an important driver in bringing wider attention to the subject. Such research has intersected with other research strands focused on both historical and contemporary green and blue spaces, and how they produce health and wellbeing (Foley and Kistemann, 2015; Gladwell et al., 2013). At the same time, and across a wider range of disciplines, there has been a parallel development in research on nature-based health and health-enabling places (Duff, 2012; Mitchell, 2013). Such research draws from a range of cognate disciplines including planning, landscape architecture, psychology, anthropology and public health (Groenewegen et al., 2012; Hartig et al., 2014).

The aim of this book is to exploit these linked interests and develop a corpus that combines human geographical and wider multidisciplinary explorations of water in a way that maintains and deepens a specific concern for health and wellbeing. Such an approach has potential theoretical and methodological value. Theoretically, it helps deepen the place of health geographers vis-à-vis contemporary theorising of health and wellbeing (Andrews et al., 2014; Crooks et al., 2018; Hall and Wilton, 2017; Kearns, 1993). Methodologically, it reflects exciting existing quantitative and qualitative approaches to inquiry, and develops these in innovative ways (Bell et al., 2015). In part, the collection draws on recent therapeutic landscapes research, but equally uses wider multidisciplinary literature to explore how other fields of scholarship measure and value the health benefits of 'nature', in both numeric and narrative forms (Keniger et al., 2013; Menatti and Casado da Rocha, 2016). In addition, recent developments that show the value of cultural and creative approaches to, and representations of, water will broaden the appeal of the book (Brown and Humberstone, 2015).

As a text grounded in health geography, it augments recent special issues and editorials in key journals, such as *Social Science & Medicine* and *Health & Place*, that advance the idea of healthy blue space (Foley and Kistemann, 2015). Additionally, however, it assembles chapters on healthy environments and practices more generally reflecting the wider concerns of nature-based health and the established importance of green space research within the arena of public health (van den Bosch and Bird, 2017). While these are important complementary reference points with green space, the main aim of our book is to explicitly consider blue space and water-based health. The book's intent is to emphasise and promote the value of blue space thinking to a variety of health and wellbeing subjects and interventions. While other water-based research has focused on pathogenic dimensions of the subject and these will not be ignored, our collection has a more salutogenic focus with a strong emphasis on place-based promotion of health, echoing what Lea (2018) calls a '*hedonic turn*'. A further connection is that recent research in the area of therapeutic landscapes (Bell et al., 2018) emphasises the importance of a fuller understanding of what makes watery places health-enabling. This, the authors claim, can be achieved through a deeper focus on experiential, emotional and embodied geographies that also recognise the wider structural settings within which those geographies are framed and produced.

### *Relational waters*

Waters have, over a considerable time span, been identifiable both as sites of healing and as material healing substances. Those long histories of water's differential role as a healing 'product' can be found across cultures, from Roman Baths and spas, to contemporary indigenous spaces from Peru to

Polynesia (Foley, 2010; Gesler, 1992). That 'product' can be found and used in many forms, from crystal purity to muddy lumps, and is strongly associated with the fact that water itself is fluid (every pun intended) in its relational associations with other elements: chemical, social, natural, cultural. All of these relational associations can, and often do, have a health and wellbeing component. There are significant curative scientific and discursive literatures, including many from within medical history and geography, on the multiple and different utilisation of water across cultures globally, especially in baths, spas, springs and wells (Foley, 2010; Gesler, 2003; Williams, 1999). These sites have been used within diverse manifestations of health and illness – for treatment and recovery (wounds, aches, pains), for health maintenance (spa cures, different chronic conditions), for mental health and wellbeing (stress reduction, contemplation, cognitive restoration) – and within a range of spiritual and religious rituals and practices (wells, springs) (Williams, 1999, 2007). The lengthy literature on therapeutic landscapes referred to above attests to the many ways in which water has been enrolled as a provider of healing in societies both ancient and modern.

But equally, in less affluent surroundings in the Global South, essential access to water for the maintenance of human life is still difficult, emphasising the ongoing importance of clean and healthy water to infectious disease management and prevention, wherein sustainable and equitable access to water is essential to basic survival. While global agencies such as the World Health Organization (WHO) or the UN International Children's Emergency Fund (UNICEF) are centrally involved on the medical and environmental health side, access to safe and clean water remains a relational process with strong contextual, social and political overtones. Environmentally shaped networks of water supply and management are linked to climate and weather assemblages that produce too much or too little water in specific places, often at the worst possible time. This variable scarcity or abundance is a reflection of the increasing unpredictability of global water patterns due to climate change. This unpredictability affects not just supply and availability, but also the condition and form of water itself as it shifts from fresh to saline, present to absent (lake shrinkage, desertification), and frozen to flowing and back again. In terms of wider environmental risk, this uncertainty makes water a potentially dangerous and uncontrollable substance, wherein flooding, drought and extreme events all threaten human life. In addition, there is a strong argument to suggest that water will be at the heart of contestations within wider social and political contexts as in the future, whereby water's resource value might make it potentially as valuable a liquid as oil (Gandy, 2014).

In focusing on human health, the nature of water makes it an especially good fit with relational thinking. Cummins et al. (2007) summarised the differences for health geographers in shifting to a relational form of thinking that was mobile, multiple, layered, dynamic and characterised by differential understandings of bodies and places shaped by personal choice, social power relations and cultural meaning, as well as life stages. Water itself is multiple and ever shifting, even if its core chemical content is fixed; this contention equally applies to its production as a material healing element and as a connective surface on which many different inscriptions take place. If one were to look at an area of lake or offshore water across time (often recommended by health psychologists for stress reduction and attention-restoration), that water can start as a *tabula rasa*, a blank flat sheet, that over an hour, day, week or more produces multiple other surfaces, colours, flows, disruptions and odours – always relational to other near and far effects of swell, weather, human and more-than-human movements and flows.

Those surfaces and depths, peaks and troughs, reflect the different stages of human health and illness which at times seem incurable but may also be temporary, even reversible. The invertibility of water, especially in seas, rivers and lakes is one of its defining characteristics; landscape becomes waterscape, fixed surfaces are rendered fluid, tides rise and fall (Ryan, 2012). This same fluidity applies to its use for health and wellbeing purposes. The idea of water as a relational and performative force and space reflects wider theoretical writing which is also slowly seeping into health geographies, especially in relation to post-humanist, non-representational and assemblage thinking (Andrews et al., 2014; Duff, 2014; Hall and Wilton, 2017). Founded on the enacted relations between people and place, this contemporary writing acknowledges a debt to early humanist geographies by Tuan (1977), Relph (1976) and Seamon (1980), within which water – as performative and (inter/en) active space – becomes the setting for *'affective becomings'*. The natural extension of Wilson's *biophilia* to our more focused *hydrophilia* seems to us a logical leap. As discussed in this book, hydrophilic theoretical concerns with embodiment, emotion and experience recognise that theoretical-empirical link, but equally recognise that those potentially transformative healthy interactions are always framed by environments, settings and contexts that are also already emergent, mobile and contingent.

### Structures and themes

Thematically, the book is organised into three parts which will be more fully introduced below with a brief description of the content of individual chapters. The three parts consider, in turn, interdisciplinary perspectives, the experience of health in blue space and finally, the value of blue space to human health and flourishing. Each of these parts reflects recent theoretical debates across health geography, in turn, informed by wider literature drawn from human geography, cultural theory and health-related realms such as public health, anthropology and psychology. In drawing carefully selected but diverse cross-disciplinary scholarship on water into a health geography corpus, we make explicit the 'spatial turns' in those subjects. The shared concerns with health, place and space will, we hope, extend in turn an awareness of the potential of fruitful collaborations with health geographers in the future.

In bringing in a concern for *interdisciplinary perspectives*, one of the core tasks of the book is to provide a coherent orientation and framework for readers from different disciplines. From the start, we acknowledge that when it comes to the relationships between health and place, scholars working within other disciplines have had plenty to say on their own terms. We make no exclusive claim to that space but rather emphasise the potential for hydrophilia and wellbeing to act as an enhanced connecting concept. One has only to look at subjects like environmental psychology, health promotion/education, public health, landscape architecture, building design and medical humanities to come across similarly inspired work (Frumkin, 2003; van den Bosch and Bird, 2017). Indeed, several of the authors in this text might specifically identify with those subjects (Atkinson and Hunt, 2019). An associated theme developed in this first section of the book is the relationship between human and more-than-human worlds, by which we mean not just the environment but the wider living dimensions of that more-than-human world. Water itself is, of course, more-than-human, but it is affected in complex ways by its relations with humans – ditto the different flora and fauna which live in water. Those relationships are evident in a number of ways; for example, the different food chains of which humans are part or resource management and wider blue economy discussions that have implicit links to human health and flourishing, especially around healthy diets, sustainable fishing activity and the potential of new marine biomedicine (Winder and LeHeron, 2017). These economic and sustainability factors are important inclusions in our consideration, and the explicit commodification of water for health/healing has a long history which continues offshore as well as inland. Finally, in considering more-than-human health, the health of water itself is continually compromised by the actions of humans – especially in relation to chemical pollution, microbial contamination and environmental degradation, something picked up in chapters that consider more fully indigenous knowledge.

There is an increasing interest in the *lived experiences* of health within blue spaces incorporating aspects of embodiment and emotion (Foley, 2017). As noted by Duff (2012), we know in broad terms that certain places enable health; we just do not know enough about how that process of enablement works. In developing recent special issues of journals and wider green/blue space health scholarship (Pearce et al., 2016), a second core theme of the book is the question of how health is experienced in place and across time and life-courses. This theme will be developed by extending accounts and narratives from a range of different blue space cultures including canals, islands and coastal populations. The chapters consider the multiple contacts between human/non-humans with water as embodied experiences. Such experiences emerge as a set of sensescapes (Bell et al., 2017) that are in turn visual (aesthetic and representational), haptic (touch, immersion, flow), sonic (trickle to roar, attention-restoration), gustatory (fresh/healthy to musty/contaminated) and olefactory (healthy and unhealthy smells). The very different and varied responses associated with the senses emphasise the multiple material/physical bodies involved; bodies with differential capacities; and bodies of all shapes, sizes, shades and conditions. As a reverse process, water relates to bodies in multiple ways that are relationally formed through the precise nature of the body-water engagement. In this book, a very direct interest in how this works for health, healing and wellbeing – as proxies for illness, treatment and positive management and maintenance – is central.

As broad advocates of a biopsychosocial model of health (though with newer relational and emplaced angles), we suggest that embodied emotional responses emerge within an affective/experiential continuum that ranges from very direct and physical contact to the more intangible impacts of water on identity-memory. Any relationally framed discussion of affective responses to water as a healing object must acknowledge its double-edged nature; just as there is hydrophilia, so there is hydrophobia. As well as its capacities to enable health, water can be and is a wounding and health-endangering object: one that is variable across time and place (Collins and Kearns, 2007). The experiential chapters in the book draw from the different relational actions and activities in which human-water engagements take place. Different bodies interact differently across life-courses, linked to family histories, mobilities associated with life events including work, risk events, exposure and access. These chapters provide valuable specific experiential accounts of health-related interactions by, on and in water around leisure and sport that link with other geographies and also to the work of environmental psychologists (White et al., 2010).

As a final key theme, we consider healthy blue spaces in the context of health and environmental inequalities and injustices. Evidence and activism on environmental justice and links with health inequalities have been developing for decades, and have great pertinence to hydrophilia. Blue environments and water more generally have the potential for universal benefit, but also have the potential for exclusion of specific groups or individuals, whether through economic, political, cultural or historic processes. There is a genuine interest in the book in exploring how robust and rigorous evidence bases can be developed through innovative and interesting methodological approaches that are more-than-quantitative. However, we also recognise the value of mixed numeric/narrative evidence; valuing blue space reflects the responsibility of critical health geographers to inform and be informed by public health politics/policy and consider spatial inequities in health outcomes across a range of different jurisdictions (Pearce et al., 2016; Philo, 2016). This section considers both benefits and risks associated with blue space including access to healthy water in the Global South and the deeper implications of climate change in terms of impacts on livelihoods and indigenous cultures. At the heart of this section, but also an important thematic aspect of the entire book, are critical reflections on the structures within which bodies experience healthy blue space. Such blue spaces are produced by elements of social and cultural politics, economics and ownership, aspects of power and inclusion/exclusion as well as wider contexts/environments in which water itself sits or moves through.

## Part B: chapter content and structures

### *Interdisciplinary perspectives*

Our book begins with a section focused on interdisciplinary perspectives on water and health, and Veronica Strang's opening chapter is an appropriate starting point. As an anthropologist with a record of engaged writing on the multiple meanings of water, her chapter contrasts cultural meanings of water – the substance of life – as a site of wellbeing within UK and indigenous Australian settings. In so doing, she explores concepts of health as a matter of orderly systemic flows within the body, and between the body and the wider material environment and assemblages of water, that reflect traditional medicine's understandings of body energies. She identifies animation as especially apposite to human health and observes that circulation in both bodies and places only really becomes noticeable when the flow is too weak or too strong. The first part of the chapter draws from ethnographic studies of the River Stour in Southern England that picks out three relevant themes. First, local perceptions of clean water are discussed as an essential hydrator, incorporating phenomenological connections around cleansing and detoxification alongside perceptions of dirty water affecting swimmers with waterborne diseases. A second theme documents the restorative and meditative qualities of water for mental health and wellness and notes how stillness and wildness emerge as oppositional materialisations of the affective nature of water. In that opposition, a balance emerges as a way of 'maintaining order'. Finally, a significant spiritual element is located within the landscape: this emerges through flows/seepages and is performed through rituals and built material landscapes. These different accounts are reflected in the second half of the chapter, which deals with a quite radically different cultural, social and environmental contexts. This entailed research in the wetlands and savannahs of the Mitchell River in North Queensland, primarily with Aboriginal Kunjen language speakers. Older colonial narratives noted (within a mostly 'barbarist' account) the health of indigenous residents as measurable by good teeth. The symbolic importance in local cultures of water emerged in the form of the 'rainbow servant' as the giver of life, and sentient presence in 'country', a phenomenologically rich blurring of person/place identity and time. Such accounts reflect the 'hypersea' metaphor of something deep underneath that bubbles up in us, via the amniotic fluid of a pregnancy, or the ways in which life-courses are understood hydro-theologically, as relational becomings, that flow out of and then back into the great artesian basins from which they emerge. The metaphor of the rainbow was also the basis of rituals, some of which involve immersion in sacred waters as a form of health transformation. Places too are transformed and dry out or flood in response to disrespectful transgressions of ancestral Law, with industrial pollution as a classic example, blocking the flow and despoiling the water as in the UK.

As writers coming from tourism and leisure geographies, Chapter 3 by Sean Gammon and David Jarratt clearly identifies how those subjects work closely with understandings of blue space as settings for health and wellbeing. In their historical introduction to the tourism and leisure components of coastal blue space, its established function as a liminal setting in both time and space is central to what they term a 'blue space-health nexus', within which leisure is identified as having a crucial intervening role. They discuss the meaning of the seaside, especially to Northern English workers identified close links with industry, and the increasing democratisation of the seaside in the 19th century via regular invasions of working-class communities coming for their annual timetabled holiday weeks. Gammon and Jarratt built around the importance of that holiday as a space of rest and recovery that reflected aspects of attention-restoration as well as a way of keeping workers happy and healthy, with a clear secondary intent of maintaining flagging productivity. But the authors also reflect on how the wider idea of escape or 'time out' has been and always will be important and that there was a heightened propensity for this in the 'spaces of difference' provided by the seaside. They develop this idea in relation to recent blue space work to identify a relational subject in place wherein the transmission of health benefits is in turn shaped by the receiver. In seeing that as akin to the transmission of a radio receiver, it speaks to the relational subject working within a blue space that they tune in to and out of. This idea of a dial controlling the input/output of nature can be linked to wider writing on stillness and the meditative potential of blue space (Bissell, 2011; Conradson, 2005; Smith, 2004). Gammon and Jarratt relate this effectively to wider concerns within wellbeing work linked to flow and what they suggest are important 'savouring' dimensions of the seaside. These observations are where overlaps with later chapters on value also emerge, characterised as, in turn, intrinsic, extrinsic and eudaimonic. Equally they identify the idea of 'seasidedness' as an orientation and draw to the coast identified in other historical and contemporary settings (Corbin, 1994; Kearns and Collins, 2016; Lenček and Bosker, 1998; Walker, 2016), albeit one that is always shaped by aspects of class, wealth and an autonomous leisure time.

As an active account, Chapter 4 provides a rich emotional and embodied auto-ethnographic account of sailing in the Auckland's Hauraki Gulf in New Zealand. Brown identifies this as an example of thalassophilic writing which aligns well with the book's intent and content. The account describes an embodied deep blue mapping for wellbeing, expressed in intensive engagements with the ocean through coastal sailing. He identifies such blue settings as spaces of learning and caring for an environment that has its own health concerns. Yet the chapter, especially in its auto-ethnographic quotes, articulates contested experience and embodied elements that place engagements with nature directly as a balance to the attention demands of a techno-connected world. Being in blue space on a boat can depend on a good GPS device for navigation, yet equally allows one to escape the tyranny of a mobile phone signal. These new definitions of wilderness – the place with no mobile phone signal – reflect earlier writing on wilderness as therapeutic settings, though this chapter extends that from mountains to the sea (Palka, 1999). For Brown, being at sea is linked to one fundamental building block of relational geographies, the family, and to aspects of work-life balance. The chapter shows how these aspects are balanced out in a relational way, wherein everyday and special places/practices become intertwined in this account. It also invokes variable depths of experience which introduce a sense of micro-embodied traces of space. Brown's account of his own red blood spattering the deck from a small accident, multicoloured seasickness and the regular mention of white foam all attest to the notion of palettes of place (Foley, 2018). Equally those micro-experiential moments are identified as affective encounters that are also material spaces of work that require attunement and skill and respect for working with the water for health and wellbeing. Brown also characterises these moments of 'discovery' as speaking to a non-representational in situ experience that is always emergent and contingent on place/time/encounter. The act of throwing oneself in or opening oneself up to an encounter with the water or the boat is often the starting point for an affective trigger memory that may incorporate other feelings such as fear (Kearns, 2018). He also notes the importance of an ethic of care and responsibility that reflects several other chapters in this volume, from experiential encounters in the Global North to cultural and social encounters at a fuller global level.

In Chapter 5, Katherine Phillips and Antony Lyons extend that concern for a dual/shared blue care, drawing from their work across a range of "hydro-citizenship" projects. They especially emphasise agendas in which a committed group of hydro-citizens co-produces health as a communal and sustainable resource within blue space. The chapter documents a case study of holistic ecological interventions within the River Churn catchment in the English Cotswolds, echoing Chapter 2 but from a different angle incorporating artistic interventions. Using WILD (Water and Integrated Local Delivery) approaches, the projects integrated individual health and wellbeing into care of a riparian environment. These initiatives led to the development of multisensory approaches to those acts of care, utilising 'minding' as a relational term that considered both a 'caring for' and a 'paying attention to' that helps frame the work as both an emplaced and embodied social/political action. For the participants, this is reflected in wider environmental volunteering research (Koss, 2010); the sense of purpose involved in the activities becomes an essential component of health promotion, echoing Antonovsky's theory of salutogenesis (Antonovsky, 1979). The wider health dimensions and specifically value of the volunteered hydro-citizenship work are referenced against a growing interest, from a policy perspective, in the potential of 'social prescribing'. Fortunately, Phillips and Lyons consider such assumptions critically and emphasise instead links between minding and mindfulness. In carrying out a combined act of environmental minding, they prefer the term 'eco-social healing', which they in turn develop to link to a more mindful socio-ecological policy and embed this in activities around the WILD initiative. Their work is valuable in providing explicit links to public health policy and a move towards 'making policy human' by making shared blue care an essential exemplar. For policy to work effectively, they also identified the importance of trust in shared ethics of care and provide intriguing links to other eco-care chapters as well.

### Experiential accounts

The second part of the book focuses on differential aspects of experiencing health in blue space. Karolina Doughty provides a valuable introductory piece to the 'therapeutic blue' when reflecting on the lengthy global histories of the use of water for health and wellbeing. That historical dimension is often forgotten in contemporary hydrophilic writing, but it provides an important globalised cultural grounding that contextualises where key terms like spas, springs and bathing places originated. She develops the chapter to critically consider what health geographers can bring to such studies, including alternative ways to value such spaces, but also that relational recognition of a shifting set of inputs and outputs across time and space. She characterises these as *'experience-scapes in the therapeutic blue'* that are explicitly related to both historical contexts and contemporary life-courses. She also responds to the need, as noted by Sarah Atkinson and others, to consider those individual experiences in relation to wider society to provide useful links back to shared longitudinal experiences of health in water. She provides a rare non-Anglophone example from Sweden on historical linkages between early curative narratives and more contemporary understandings of health promotion. From a contemporary theoretical perspective, Chapter 6 also develops the idea of experience-scapes smoothly into relational geographies that are multiple, mobile and operate across scales. The specific emphasis on the multiple chimes well with recent reviews of therapeutic landscape writing (Bell et al., 2018) that argue for an enhanced focus on diversity of experience and outcomes. In picking out work that promotes an affective, emotional and embodied approach, there is a strong development of those elements in the chapter, something that is a key theme running through the book as a whole.

In keeping with that idea of experience-scapes in the therapeutic blue, Chapter 7 by Easkey Britton documents surfing as an act of health and wellbeing in two quite contrasting settings, Ireland and Papua New Guinea (PNG). In bringing in more specifically performative and immersive examples, the surfer's experience is presented as a set of voices from within the waves, actively shaped by the settings and the individuals involved. This perspective provides an explicit global/local edge with contrasting and yet somehow common cultural relationships with both human and place-related health. In focusing on immersion, she emphasises the embodied experience, through both her own auto-ethnographic writing and other voices, of what it feels like to surf, to interact with variable wave heights and energies, to experience both the highs and lows of experiences that are simultaneously health-enabling and risky. In drawing from older Irish cultural histories, she develops the chapter conceptually around the idea of the borderlands of surf/zone that recognises the liminality of both physical and cultural space within quite different contexts. That extends to a personal relational immersion, via her own upbringing in one of Ireland's pioneering surfing families that becomes over time a therapeutic accretion. That personal therapeutic accretion reflects Brown's earlier reference to the work of interacting healthily with blue space, here too linked to expertise, respect and skill as important and under-explored dimensions of wellbeing. A further liminal positionality, as a woman in a generally very male world, is specifically linked in the chapter to her history as a big wave surfer and through the empowerment of women surfers in unlikely settings like PNG and Ireland. The chapter also provides useful empirical material on the more theoretical work around more-than-representational theory (Andrews, 2018) that recognises those new directions but retains a regard for the representational. In surfing, there are very specific place and practice languages taking place and it is striking, even in a country as polyglot as PNG, that similar *tok pisin* (pidgin) terms for surfing, *kisim solwara*, *kisim sea* or *katim solwara*, are used around the country. At heart, the chapter provides a kinetic account that brings experience and feeling alive, yet recognises the contested sides of danger and potential serious injury as well, both in the joy and inherent fear in the act of surfing across a range of expertise

A very different form of hydrophilic experience is recounted by Allison Williams, Ashleigh Patterson and Morgan Parnell in their discussion of a yoga retreat in the Bahamas. The chapter, based on participant observation and interviews with practitioner attendees, develops earlier writing on globalised therapeutic landscapes and yoga (Hoyez, 2007) within a specifically blue space setting. The chapter also draws from wider wellbeing writing and health tourism but develops this through documented experiential wellness practices and their associated micro-geographies. For those who can afford it, the seeking out of experiences and the associated choice of mode location are important in bringing yoga practitioners towards a contemporary (blue) wellness setting where they can enact hydrophilia through a specific wellbeing practice. Within that space, and reflecting equally the very particular nature of yogic practice, itself complex, differential and contested, raises questions around the 'authenticity' of the experiences and a slight snobbery associated with the different communities of practice which patronise the resort. This concern with authenticity is a common trope within writing on wellness tourism (Smith and Kelly, 2006; Smith and Pucsckó, 2009), and is especially evident in discussions around the spiritual dimension linked to yoga and the wider places and settings within which it is practised. The location described in the chapter could certainly be described as exotic and therein lies some of the contestation, though the comparisons with what might be termed everyday yogic spaces are discussed critically. The chapter provides an account of gendered and experiential response to both the place and the wellness practices in that place. Having the correct atmosphere for yoga is, it is argued, enhanced in nature via direct exposure as an experiential health imbrication (Conradson, 2005). This is in turn shaped by place, as noted in McCormack's (2018) consideration of affective/atmospheric geographies and the underestimated role that place plays in personal affective interactions, reflected across writing on experience- and sense-scapes. The yoga retreat provided opportunities for volunteers alongside sometimes affluent visitor/practitioners and some tensions emerge in the accounts of these different positionalities. Finally, there is an important discussion on external environmental risk and the role of an *'unpredictable blue'*; especially given the resort has been damaged several times in recent hurricane seasons. In those very place characteristics that draw people to the resort for health and wellness, the harnessing of the healing power of the blue is never fully guaranteed and often works the opposite way.

Chapter 9 by Hannah Pitt documents how canals are used for health and wellbeing within a number of deprived communities in the UK. As new examples of blue spaces, she lists the contested representations and differential palettes of what she terms *'an unpromising blue'*. She is also keen to emphasise the hydrophobic alongside the hydrophilic dimension. Yet through the work itself, based around initiatives associated with the Canal and River Trust (CRT), Pitt identifies the canals as potentially transformative spaces for normally excluded and marginalised populations that might be pejoratively termed as equally unpromising. The chapter describes how representative BAME (Black and Minority Ethnic) subjects, a group with initially hydrophobic perceptions of canals, were exposed to initiatives on the canals via the work of the CRT. The experience of these initiatives shifted their perceptions significantly and led to their identification of such spaces as potentially inclusive and positive in terms of health and wellbeing. The chapter is indicative of a new wave of work into different types and shades of blue and wider assemblages of shadings and meanings. In an initiative around the use and care of blue space, we can identify another public/social intervention with additional health benefits. Equally, the mix of human and more-than-human vision is in turn socially and critically engaged with marginalised experiential voices, which speak to a call for narratives from multiple sources and settings (Bell et al., 2018).

## Blue health inequality and environmental justice

The final set of chapters in the book considers important policy dimensions around how blue space and health relationships exist within wider discourse of environmental and health inequality and injustice. Chapter 10 by Sarah Bell, Rebecca Lovell, Julie Hollenbeck, Mat White and Michael Depledge uses three case studies from the US and UK that bring attention to differential user experiences – incorporating emotional and embodied components – that consider both the philic and phobic dimensions of water for health. In linking this to how we might value blue space, the chapter uses a realistic notion of balance around potential gains/losses that echo those same dimensions. The initial section of the chapter reflects on specific ways in which different 'sense-scapes' emerge for different populations, with a particular emphasis on the blue. It focuses on research in the UK with an under-regarded and excluded group, people with sight impairment. In exploring how nature is sensed by this group the research uncovers rich mediated experiences that are contingent on level of impairment and individual capacities and resources, reflecting a balancing out of the individual/social. The second section is based on a case study from Florida of historically unequal access to beaches for African-Americans. Here the knock-on effects of disallowing a socially produced right to water are tracked across an ethno-racial experience but have longer-term intergenerational implications as well in relation to water knowledge and risk (Wiltse, 2010). This negative experience stands in stark contrast to any therapeutic accretions that might occur in such settings, chipping them away instead of building them up. While focused on distinct African-American experiences, those access/exclusion elements reflect coastal geographies elsewhere where exclusion is equally bound up in cost, ownership, legal and bounded dimensions of a space that is globally valued (Gesler, 1992). The final section of the chapter extends specific experiences of exclusion by disability and race and scales it out to more global scale. Taking into account risks to wider coastal communities from climate change and environmental degradation, a relational geography operating across multiple settings and scales results in a visible devaluing of specific groups and communities who live in such places. That threat to the community-level benefits of blue space requires a more inclusive vision of health and wellbeing that governmental and private planning should and must engage with.

In considering aspects of inequality and injustice of water at a global scale, Carmen Anthonj and Timo Falkenberg broaden the scale out again, well beyond the coast, to consider the specific challenges of environmental health in the Global South, utilising the concept of '*thirst world*'. While earlier chapters have focused on individual or community scales, taking a bigger picture to look at a continental level, in this case Africa provides a useful corrective and challenges comfortable assumptions about water in the Global North. Reference in Chapter 10 to the idea of a right to water does not translate easily in thirst world. In this world, there is a relational value linked to the basic presence/absence of clean and drinkable water that is also mobile around reliability, affordability and quality that continues to shape global health inequity. The chapter also provides a useful medical geography perspective on blue space that is a (literal) world away from leisure and wellness, where the contested outcomes in terms of the balance between health-enabling and health-endangering aspects are considerably starker. In addition, the sometimes precious concerns around palettes of blue space are overturned. The luxury of pristine clear water is rarely available, yet muddy palettes abound in thirst world, where as long as they are healthy they are perfectly acceptable to the essential maintenance of human health and flourishing.

While a starkly contrasting example, Chapter 12 on wild swimming as a healthy blue practice provides another take on aspects of community equity and everyday practice that explores the health value of everyday immersion in blue space. Taking as its starting point the phenomenon of 'wild swimming' in the UK especially, Sarah Atkinson examines both swimmers and their preferred outdoor swimming spots (lidos, lakes, rivers, reservoirs, beaches) as informal communally produced sites of blue health. While the term itself is noted as having a whiff of commodification, it does also tap in to a craving for wildness and wild places within contemporary lifestyles that tallies with therapeutic landscapes writing on retreat, stillness and immersion (Foley, 2017). In exploring access to swimming spots there are certainly echoes of earlier chapters' identification of blue spaces as being part of a public good and access to certain swimming locations remains a contested experience for wild swimmers, linked to the growing reach of risk narratives around health and safety as well as emplaced privacy issues. The former acts as a sort of undertow across the whole book, where traditionally open public blue spaces are increasingly fenced off, both literally and metaphorically, from public discovery and exploration. In Chapter 12, Atkinson deftly challenges the perceived individualised aspects of wild swimming with the wider social and spatial contexts in which it has emerged as a branch of informal public health and engagement with place. The world of wild swimming is also critically examined in the chapter. For some, it can be perceived as a distinctly middle-class experience though this seems a slightly harsh and untrue judgement. Indeed, the author outlines and effectively contrasts older histories of public bathing as a contrast to the more privatised contemporary model. As old public pools are closed or run-down, new private clubs and restored 'luxury' baths are replacing them. In using pools provided by nature such shifts are challenged and the development of free spaces for swimmers in the outdoors builds up a more inclusive swim-for-all perspective. The chapter mentions the differential distribution of the health benefits of 'self-care' activities such as wild swimming, with an important thread reflecting on blue space and gender. In particular, it highlights that wild swimming actually appears to buck the gender differential of many outdoor activities, in that female participants outnumber males (at least in some events/places). Finally, the chapter, framed within a medical humanities perspective, teases out aspects of how swimmers themselves reinterpret, through their emplaced practices, a new form of public health that develops a sense of *'unmediated authenticity'* in the bodies and minds of swimmers. In terms of valuing blue space experiences and practices, policymakers could do worse than draw from such communities of practice.

Chapter 13 by Meg Parsons takes a place-based perspective on contested water-based environments, exploring colonial v. indigenous cultural perspectives in the Bay of Plenty in the North Island (*Te Ika a Maui*) of Aotearoa New Zealand. Here health emerges as a set of balances between human and more-than-human worlds, cultures and nature and local and global perspectives, wherein water produces health in complex relational ways. The chapter outlines the ways in particular in which colonial authorities set out to tame the waters in this coastal zone and contrasts these with indigenous Māori understandings of healthy place that reflect the multivalency identified by Strang in Northern Queensland. In essence, the chapter explores contested values and valuings of a liminal coastal landscape, where the balancing of perspectives on how land should be managed, assets and human life protected and health produced via diet and shared resources have ebbed and flowed. That ebbing and flowing is made explicit through different maps that show not only how rivers and tides were controlled, but also how different perceptions of place reflect wider writing on environmental justice and shifting presentations of healthy place (Gatrell and Elliott, 2015). Reflecting studies of colonial cities like Freetown or Dunedin, where the rich lived on hills and the poor on marshy, malarial flat land, the Bay of Plenty also used representations that placed the indigenous populations as living in malodorous uncontrolled tidal riparian zones and contrasted these with the 'improvements' brought about by colonial civilisation. There was a certain irony in the presenting of the liminal muds of the coastal rivers as unhealthy, given the explicit promotion by the government of mud springs in the early 1900s. But Parsons usefully identifies the wider impacts of drainage and 'unwatering' and how this challenged indigenous understandings of water as a more-than-human fluid vein of life. As was the case with other indigenous viewpoints, the act of constriction through the use of levees or bunds potentially damages a healthy flow. They also noted that the blurred palettes of green/blue implicit in most of the other chapters are here made clearer. The act of channelization very deliberately separated out the green and blue but this attempt to bind water is, as the book title suggests, ultimately a futile act in a relational blue space that is at its best, open to the new connections that health and nature co-produce.

# References

Abram, D. (1996). *The Spell of the Sensuous.* London, Vintage.
Anderson, J. and Peters, K. (2014). *Water Worlds: Human Geographies of the Ocean.* Farnham, Ashgate.

Andrews, G. (2018). Health geographies I: The presence of hope. *Progress in Human Geography,* 42 (5), 789–798.

Andrews, G., Chen, S. and Myers, S. (2014). The 'taking place' of health and wellbeing: Towards non-representational theory. *Social Science and Medicine,* 108, 210–222.

Antonovsky, A. (1979). *Health, Stress and Coping.* San Francisco, Jossey Bass.

Atkinson, S. and Hunt, R. (2019). *Geohumanities and Health.* London, Springer.

Bartram, J. (ed) (2017). *Routledge Handbook of Water and Health.* London, Routledge.

Bell, S., Foley, R., Houghton, F., Maddrell, A. and Williams, A. (2018). From therapeutic landscapes to healthy spaces, places, and practices: A scoping review. *Social Science & Medicine,* 196, 123–130.

Bell, S., Phoenix, C., Lovell, R. and Wheeler, B. (2015). Using GPS and geo-narratives: A methodological approach for understanding and situating everyday green space encounters. *Area,* 47 (1), 88–96.

Bell, S.L., Wheeler, B.W., Phoenix, C. (2017). Using geonarratives to explore the diverse temporalities of therapeutic landscapes: Perspectives from 'Green' and 'Blue' settings. *Annals of the American Association of Geographers,* 107, 93–108.

Bissell, D. (2011). Thinking habits for uncertain subjects: Movement, stillness, susceptibility. *Environment and Planning A,* 43, 2649–2665.

Brown, M. and Humberstone, B. (eds) (2015). *Introduction, Seascapes: Shaped by the Sea.* Ashgate, Farnham.

Collins, D. and Kearns, R. (2007). Ambiguous landscapes: Sun, risk and recreation on New Zealand beaches. In, A. Williams (ed), *Therapeutic Landscapes.* Ashgate, Aldershot, 15–32.

Conradson, D. (2005). Landscape, care and the relational self: Therapeutic encounters in rural England. *Health & Place,* 11 (4), 337–348.

Corbin, A. (1994). *The Lure of the Sea: The Discovery of the Seaside in the Western World 1750–1840.* Cambridge, Polity Press.

Crooks, V., Andrews, G. and Pearce, J. (eds) (2018). *Routledge Handbook of Health Geography.* London, Routledge.

Cummins, S., Curtis, S., Diez-Roux, A. and Macintyre, S. (2007). Understanding and representing 'place' in health research: A relational approach. *Social Science & Medicine,* 65, 1825–1838.

Duff, C. (2012). Exploring the role of 'enabling places' in promoting recovery from mental illness. A qualitative test of a relational model. *Health & Place,* 18, 1388–1395.

Duff, C. (2014). *Assemblages of Health. Deleuze's Empiricism and the Ethology of Life.* New York, Springer.

Foley, R. (2010). *Healing Waters: Therapeutic Landscapes in Historic and Contemporary Ireland.* Farnham, Ashgate.

Foley, R. (2017). Swimming as an accretive practice in healthy blue space. *Emotion, Space and Society,* 22, 43–51.

Foley, R. (2018). Palettes of place: Green/blue spaces and health. In, V. Crooks, Andrews, G. and Pearce, J. (eds), *Routledge Handbook of Health Geography.* London, Routledge, 251–258.

Foley, R. and Kistemann, T. (2015). Blue space geographies: Enabling health in place. Introduction to special issue on healthy blue space. *Health & Place,* 35, 157–165.

Frumkin, H. (2003). Healthy places: Exploring the evidence. *American Journal of Public Health,* 93 (9), 1451–1456.

Gandy, M. (2014). *The Fabric of Space: Water, Modernity, and the Urban Imagination.* London, MIT Press.

Gatrell, A.C. and Elliott, S.J. (2015). *Geographies of Health: An Introduction* (3rd ed.). Chichester, Wiley Blackwell.

Gesler, W. (1992).Therapeutic landscapes: Medical issues in light of the new cultural geography. *Social Science & Medicine,* 34 (7), 735–746.

Gesler, W.M. (2003). *Healing Places.* Lanham, MD, Rowman and Littlefield Publishers, Inc.

Gladwell, V.F., Brown, D.K., Wood, C., Sandercock, G.R. and Barton, J.L. (2013). The great outdoors: How a green exercise environment can benefit all. *Extreme Physiology & Medicine,* 2 (3), 1–7.

Groenewegen, P., van den Berg, A., Maas, J., Verheij, R. and de Vries, S. (2012). Is a green residential environment better for health? If so, why? *Annals of the Association of American Geographers,* 102 (5), 996–1003.

Hall, E. and Wilton, R. (2017). Towards a relational geography of disability. *Progress in Human Geography,* 41 (6), 727–744.

Hartig, T., Mitchell, R., de Vries, S. and Frumkin, H. (2014). Nature and health. *Annual Review of Public Health,* 35, 207–228.

Hoyez, A. (2007). From Rishikesh to Yogaville: The globalization of therapeutic landscapes. In, A. Williams (ed), *Therapeutic Landscapes.* Aldershot, Ashgate, 49–64.

Kearns, R. (1993). Place and space: Towards a reformed medical geography. *Professional Geographer,* 45 (2), 139–147.

Kearns, R. (2018). Rituals and performance crossing the line: All at sea with King Neptune mid-pacific. In, M. Brown and Peters, K. (eds), *Living with the Sea. Knowledge, Awareness and Action.* London, Routledge, 213–226.

Kearns, R. and Collins, D. (2016). Aotearoa's archipelago: Re-imagining New Zealand's island geographies. *New Zealand Geographer,* 72 (3), 165–168.

Keniger, L., Gaston, K.J., Irvine, K. and Fuller, R. (2013). What are the benefits of interacting with nature? *International Journal of Environmental Research and Public Health,* 10, 913–935.

Koss, R.S. (2010). Volunteer health and emotional wellbeing in marine protected areas. *Ocean & Coastal Management,* 53 (8), 447–453.

Lea, J. (2018). Non-representational theory and health geographies. In, V. Crooks, Andrews, G. and Pearce, J. (eds), *Routledge Handbook of Health Geography.* London, Routledge, 144–152.

Lencek, L. and Bosker, G. (1998). *The Beach: The History of Paradise on Earth.* New York, Viking.

McCormack, D. (2018). *Atmospheric Things: On the Allure of Elemental Envelopment (Elements).* Durham, NC, Duke University Press.

Menatti, L. and Casado da Rocha, A. (2016). Landscape and health: Connecting psychology, aesthetics, and philosophy through the concept of affordance. *Frontiers in Psychology,* 7, 571.

Mitchell, R. (2013). Is physical activity in natural environments better for mental health than physical activity in other environments? *Social Science & Medicine,* 91, 130–134.

Palka, E. (1999). Accessible wilderness as a therapeutic landscape: Experiencing the nature of Denali National Park. In, A. Williams (ed), *Therapeutic Landscapes: The Dynamic between Place and Wellness.* Lanham, University Press of America, 29–51.

Pearce, J., Shortt, N., Rind, E. and Mitchell, R. (2016). Life course, green space and health: Incorporating place into life course epidemiology. *International Journal of Environmental Research and Public Health,* 13 (3), 331–342.

Philo, C. (2018). 'Healthy debate' and 'healthy ferment': Medical and health geographies. *Progress in Human Geography.* doi:10.1177/0309132516678343

Relph, E. (1976). *Place and Placelessness.* London: Pion.

Ryan, A. (2012). *Where Land Meets Sea: Coastal Explorations of Landscape, Representation and Spatial Experience.* Farnham, Ashgate.

Seamon, D. (1980). Body-subject, time-space routines, and place-ballets. In, A. Buttimer and Seamon, D. (eds), *The Human Experience of Space and Place.* New York, St. Martin's Press, 148–165.

Smith, A. (2004). A Maori sense of place? – Taranaki Waiata Tangi and feelings for place. *New Zealand Geographer,* 60 (1), 12–17.

Smith, M. and Kelly, C. (2006). Wellness tourism. *Tourism Recreation Research,* 31 (1), 1–4.

Smith, M. and Pucsckó, L. (2009). *Health and Wellness Tourism.* Oxford, Butterworth-Heinemann.

Tuan, Y. (1977). *Space and Place: The Perspective of Experience.* Minneapolis, University of Minnesota Press.

van den Bosch, M. and Bird, W. (eds) (2017). *Landscape and Public Health.* Oxford: Oxford University Press.

Walker, A. (2016). *Man vs Ocean: One Man's Journey to Swim the Seven Seas.* London, John Blake Books.

White, M., Smith, A., Humphryes, K., Pahl, S., Snelling, D. and Depledge, M. (2010). Blue space: The importance of water for preference, affect, and restorativeness ratings of natural and built scenes. *Journal of Environmental Psychology,* 30 (4), 482–493.

Williams, A. (ed) (1999). *Therapeutic Landscapes: The Dynamic between Place and Wellness.* Lanham, MD, University Press of America.

Williams, A (ed) (2007). *Therapeutic Landscapes.* Farnham, Ashgate.

Wiltse, J. (2010). *Contested Waters. A Social History of Swimming Pools in America.* Chapel-Hill, University of North Carolina Press.

Winder, G.M. and Le Heron R. (2017). Assembling a blue economy moment? Geographic engagement with globalizing biological–economic relations in multi-use marine environments. *Dialogues in Human Geography,* 7 (1), 3–26.

# Part I

# Interdisciplinary perspectives on water and health

# 2 The meaning of water to health

## Antipodean perspectives on the 'substance of life'

*Veronica Strang*

## Introduction

There is no 'health', either for humans or for environments, without water. To return to the roots of the word: 'hale' or 'whole' is to understand at once that the integrity of any organic system – its wholeness, haleness, heartiness, health – depends upon hydration. This is as true of the most microscopic organism as it is of whole ecosystems. Water flows through and connects every scale of health and, because we 'think with water', we conceptualise these flows in both material and non-material terms (Bachelard, 1983; Chen et al., 2013). Water connects not only micro and macro physical systems but also allows us to imagine – both positive and negative – flows of matter, as well as ideas, knowledge and the spirit, over multiple spatialities and temporalities. This chapter considers how, via what Bourdieu called 'scheme transfers' (1977), water enables coherent transpositions of ideas about health across physical and conceptual domains.

Water's material properties are consistent at all scales: its molecular composition means that, whether in the domestic kitchen or in planetary cycles, it responds to environmental conditions, transforming from ice to fluid to mist, and condensing back again. Its unique physical structure allows it to bind with, and thus to carry, a vast range of other substances, which may be benign or otherwise to the organisms and systems through which it flows. It can therefore bring vital hydration and nutrients, carry away waste, or inject poison and pollution. Water supports all of the body's chemical transformations, and also its mental processes, in that the electrical firing of neurons is enabled by water and its molecular connectivities. Above all water flows, sometimes with weight and force, acting upon bodies and environments. In this way, its properties and behaviours allow us to compose concepts of flow, movement and transformation, and to consider hydration and the ingress and egress of matter that supports or detracts from health at all levels.

Ideas about the proper flows of water and matter provide a vision of order. Health is an ultimate measure of order that, as Cohn observed, rests on a concept of balance (1997). However, this is not merely a question of stability: health is processual, depending upon flows that, to provide positive wellbeing, must be sufficient and timely, but not excessive. Slowing to relative stillness may be beneficial, and apparently still bodies of water actively encourage meditative reflection, but for health, or sustainability, in any organic system, there must be some flow and movement (Féaux de la Croix, 2011). Total immobility is at best a form of holding and waiting, at worst stagnation and non-life. Health requires constant and orderly flows of water, matter and energy.

Such a concept of health is readily transferable to multiple scales. Thus, Edgeworth points to the dynamic energy expressed in the flow of rivers:

> Put your hand into a river, or immerse any part or your body, and you can feel these cosmic and planetary forces at work in the form of the flow or current of water ... It is ... the flow or energy of the river itself.
>
> (2014: 159)

Approaching this observation from a different direction, Atran suggested that animation is the criterion that, in Western thought, most clearly separates categories of living and nonliving kinds (1990). In cultural contexts containing beliefs about sentient ancestral landscapes and objects, this notion is extended, as may be seen in the second ethnographic example below. But whether extended or not, a focus on animation and energy helps us to see why ubiquitous beliefs about 'living water' and water as 'life itself' are not merely a recognition of water's vital hydrological necessity to all organic life forms. They also relate to its animated movement and its agentive capacities to act in and upon the world (Strang, 2014).

## Bodies of water

In 1994, the McMenamins introduced the idea of the 'hypersea'. They observed that organic life on Earth began deep in its oceans, and remained there for eons, before making its way onto land. Even when they had successfully colonised terra firma, living beings retained their absolute dependence on water and continued to be composed largely *of* water. The percentage of water in the human body is approximately 67%, though generally slightly more in babies and slightly less in older people. In this way, the McMenamins suggest, organic life on land carries a 'hypersea' of water which flows between and connects all such life across both space and time. Human experience of these flows is direct and embodied: we ingest water and nutrients from environments, and expel wastes into their larger recycling systems. And – readily transposing concepts of flow and circulation into these systems – we recognise the same processes in other living beings and ecosystems, and their similar 'health' needs for sufficient, timely and unpolluted flows of water.

As noted elsewhere (Strang, 2004), we also transpose these ideas metaphorically into other systems, with ideas about what constitutes orderly economic circulation; appropriate movements of people, substance and identity; and flows of knowledge (see also Lakoff and Johnson, 1980). In each systemic domain, there is the potential for flows to be both positive and negative both in volume and in quality. Funds may be a life-saving monetary injection, or require laundering. Knowledge may be enlightening or corrupting. The flow of migrants into communities may be seen as enhancing or detrimental, with much depending on the capacities of host communities to co-identify with them through concepts of common substance.

This point brings us to concepts of health that are more than material in form. We are only intermittently conscious of water's flows through our bodies. The sensory relief of a thirst-quenching drink, or of emptying the bladder, might register; and impeded flows and the retention of tainted water might make themselves felt through swollen limbs, kidney stones or prostate enlargement. But unless 'normal' flows are disrupted, bodily hydration can otherwise be taken for granted. As Illich notes (1986), we are similarly oblivious to the invisible movements of water flowing through our cities, only becoming sharply aware of them when floods inundate the streets, or household taps fail to yield water supplies.

Nevertheless, people's phenomenological experiences of water in its various forms are profound, and have multiple affective and physiological effects. Such experiences are as diverse as the cultural and geographical contexts in which they occur, and yet even on opposite sides of the planet, and in markedly different environments, there is commonality in the ways that people think about water and health. A comparison of two ethnographic studies allows us to consider how (and why) core themes of meaning relating to water and health recur (Figure 2.1).

*Figure 2.1* Mill on the Stour.
Source: Author.

## On the River Stour

At the time when Thomas Hardy described the Stour Valley in Dorset, on the south coast of England, the river was central to local farming. It contained, in its 77 mile course, over 60 water mills, which, as well as providing power for milling corn and fulling cloth, bound the riparian communities together in necessary collaborations over water flows. A century later there is still a strong farming community: dairy herds in the fertile meadows by the river; some arable farming, and some sheep up on the chalk downs. But there are now far fewer small farms and more industrial-scale agricultural enterprises. Wealthy manorial holdings remain, including Henry Hoare's famous 18th-century stately home and graceful landscaped gardens at Stourhead. The pretty thatched cottages in villages along the river contain many retirees from London and are popular B&B locations. Small historic towns, such as Dorchester and Lyme Regis, are similarly attractive to tourists, as is the pebbled 'Jurassic Coast' along which it is still possible to discover dinosaur remains, as Mary Anning did in the early 1800s.

Over several years, I conducted ethnographic research along the river, exploring people's engagements with water (Strang, 2004) and interviewing a cross section of local water users, including farmers, conservationists, householders, schoolchildren, tourists and others, as well as people working for local water suppliers and government agencies. The research participants highlighted three major areas that they found particularly meaningful. The first was most directly related to health: they had clear ideas about the importance of drinking sufficient water and ensuring that this was not compromised in terms of quality. There were multiple anxieties about the effects of industrial agriculture and other forms of production on water quality, and doubts about the extent to which water suppliers and water regulators could be trusted.[1] Concerns that, a few centuries ago, might have focused on the moral purity of the spirit were expressed in terms of maintaining bodily purity and contemporary languages about 'toxins' and 'detoxification'. Standards of hygiene were framed in similarly material terms, with the cleansing capacities of water providing protection against invading germs and dirt. In a river valley prone to flooding, concerns about the potential for 'foul flooding' – the ingress of sewage, and thus the dead matter of others – into private domestic spaces elicited particular horror.

Such concerns about quality and the control of polluting flows impinged upon another powerful mode of engagement: immersion in bodies of water. Few people swam in the Stour at the time, partly because river swimming had fallen out of fashion and partly due to anxieties about water quality. Young people leapt happily into the water at a few wider reaches and millponds, but most people preferred to travel the short distance to the sea, where they felt that salt and dilution would mitigate any pollution. With intensive dairy farming in the valley, concerns about slurry (liquid animal faecal matter) and other farming wastes running into the river were high. River swimming has become more fashionable in the UK in recent years, with strong promotion of its health benefits. Yet rivers such as the Stour continue to veer between being held up by the Environment Agency (in 2011) as one of the most improved rivers in the country (Daily Echo, 2011) and reports of children being taken ill after swimming in it (in 2016).

> More than 10 children and two adults are believed to have contracted a sickness bug after being in the water at Eyebridge near Wimborne, Dorset ... Wendy Richardson, from Bearwood, said her nine-year-old daughter Lola had also been ill during school holidays after taking a dip at the site ... "It's a stunning location – the water is as clear, as clear can be, and so many people were swimming in it – I was never worried about it at all" ... Public Health England (PHE) said swimming in rivers and lakes can cause gastrointestinal infections, as well as respiratory, skin, ear and eye infections. "These open water areas can contain sewage, livestock contamination and pollution from farming and industry."
>
> (BBC, 2016)

Where there were no major concerns about pollution, most of the research participants (other than a small minority fearful of water) described immersion as a positive experience, providing a sense of weightlessness and freedom (Suedfeld et al., 1983). Highlighting the sense of relaxation and wellbeing that it engendered, several drew analogies with 'a return to the womb' to convey how it gave them a sense of being held and contained. They referred to the 'rocking' provided by water, and the 'soothing' and rhythmic sounds of waves and rivers. It is therefore unsurprising that 'recreational' activities so often involve these kinds of interactions with water and that their phenomenological and perceived therapeutic effects are integral to the success of holiday resorts and spas (Corbin, 1994; Foley, 2015).

There is wider recreational use of the Stour Valley: in particular its riparian walks, parks and gardens. Informants recounted with enthusiasm their affective responses to the visual stimuli provided by water bodies. Their descriptions suggest that the meeting of light and water, and the shimmering, numinous images that this composes are irresistible. Eyes and mind are drawn to water's compelling movements, as to the flicker of a fire. This leads people to use terms such as 'mesmerising' and 'hypnotic' to describe the power of water, not only to seize their attention, but also to evoke emotional and aesthetic responses and, critically, to provide healing and 'recreation' of the self (Kaplan, 1995; Ohly et al., 2016). The quiet water body – the lake, the gentle wave – served to induce reflection and memories, to enable meditation, to calm and to provide rest. Alternatively, the wild waterfall, or crashing shore, was described as inspiring, heartening and exciting. And, as Ingold found in exploring engagements with weather, water's wider movements are also deeply felt:

The experience of weather lies at the root of our moods and motivations; indeed it is the very temperament of our being ... Earth and sky,
then, are not components of an external environment with which the
progressively 'knowledged-up' (socialised or enculturated) body interacts. They are rather regions of the body's very existence, without which
no knowing or remembering would be possible at all.

(Ingold, 2010: 122)

This view suggests that as well as having a straightforward material relationship with health, water is equally implicated in emotional and mental
health, providing phenomenological experiences that 'maintain order' in
these dimensions of wellbeing as well. The language of mental and emotional flows readily acknowledges this perspective. Thinking may be effer-
vescent and sparkle with clarity, or it might be muddy and sluggish. Water
is used to discuss emotional ebbs and tides, at both individual and collective
levels. The heart may, as Sassoon put it, be 'filled with delight' (1909) or
flooded with grief and in either case such flows of emotion may be externally
expressed in tears.

Although spiritual issues may be less central to contemporary everyday
life, and the discursive emphasis has shifted to water's role in material and
emotional health, religious or spiritual wellbeing remains important to many
people, with holy wells continuing to provide therapeutic sites (Foley, 2010,
2011). Dorset is (for the most part) a conservative and primarily Christian
area. However, its earlier religious history and some contemporary alterna-
tives are strongly in evidence, with neopagans inspired by its Celtic wells,
proximity to Stonehenge and the presence of the notably masculine chalk
giant inscribed on the hillside at Cerne Abbas. The holy well below the latter
site was the subject of a classic early Christian takeover, during which the
generative powers of the ancient well to ensure fertility were reattributed
to St Augustine. According to a local myth (suggestive of some competition with the virile giant above), the Saint stuck his staff into the ground,
causing water to spout forth. However, the well continues to attract pagan
rather than Christian rituals: pieces of cloth are tied to the surrounding
trees, and contemporary 'fertility rites' are reportedly still performed in the
fields around the giant.

A number of the churches in the shire have stoups intended to contain
bowls of holy water. There are reports of ancient exorcisms, which required
holy water to restore health. Occasionally wellheads have also been used as
gravestones, drawing attention to a major complex of ideas about springs
and wells, fountains of life and enlightenment (Strang, 2004).

Such markers also highlight the relationship between water and the health
and integrity of local social and spiritual identity. A primarily rural area,
Dorset maintained considerable long-term demographic continuities until
relatively recently, with many extended 'blood relations' defining a common
local substance of identity. There are commensurate anxieties about the flow

of 'incomers', in particular wealthy retirees from London, who have pushed up house prices to the extent that the younger generations of local communities can no longer hope to own their own homes. There is also a more general undercurrent of concerns about 'foreigners', whose otherness is seen to have the potential to bring in ideas and practices (such as different religious views) that might disturb the order of established cultural norms (Figure 2.2).

Such ideas are manifested in St Peter's Pump at Stourhead and in the garden's famous grotto beside the lake. Here, overseen by a nymph and a classical river god, the source of the Stour (with a little judicious channelling to ensure a confluence of a number of springs) emerges into a pellucid pool. As in many such places around Europe, people 'take the waters' in the hope that it will improve their health and wellbeing in secular, physical terms, but the religious history of the site provides a powerful imaginative context. Indeed, the whole of the surrounding lakeside garden was conceived as a spiritual/life journey taking the pilgrim through an arboretum and contemplative temples, to conclude, properly, at the local Church (Figure 2.3).

The role of water in maintaining spiritual health is thus articulated in the local religious landscape and its architecture. It is reflected in the baptismal and mortuary rituals that, for many people, still mark the arrival and departure of individual souls to and from the material domain. Such rituals

*Figure 2.2* St Mary's Church with wellhead as gravestone.
Source: Author.

*Figure 2.3* Stourhead Gardens.
Source: Author.

resonate with a common view of the river as a metaphor of lifetime, which proceeds from the springs of infancy, through the rocky highlands of youth, to mellow in the broader plains of maturity before finally leaving individual identity behind and dissolving into the Great Sink of the sea. Implicit in this metaphor is a vision of subsequent evaporation and rising hydrological movement back to the beginning. Running through this inherently fluid life movement is a theme of moral health, in which clean water provides an image of purity and probity, uncontaminated by the stain of sin or moral turpitude. As in other aspects of health and water, ideas about how this can affect physical and moral wellbeing are readily transposed to environmental concerns, as illustrated, for example by Kelly's historical account of shifting evaluations of agricultural and national park landscapes in Dartmoor (2015).

In Dorset, a similar combination of material and moral concerns can be seen in local ideas about the ecological wellbeing of the river. The area supports considerable dairy production and there have been efforts to resuscitate the ancient water meadows along the Stour.[2] But modern intensification of such activities has led to problems with fertilisers and slurry making their way into the river, with commensurate impacts on water quality, due to faecal pollution and eutrophication, and on water flows impeded by the resulting excessive weed growth. There are similar concerns about the use of pesticides and herbicides by arable farmers, with fears that these are making their way into local drinking water supplies.

The area therefore contains some sharp local diversity in ideas about health, 'order' and generative power. For the farmers, healthy lifeways are represented by strong agricultural productivity and economic growth. For conservationists, industrial farming's impacts on biodiversity are a major loss of environmental health. For many local groups, the effects of farming chemicals on food and water quality raise concerns about their own bodily health and wellbeing. There are thus some tensions between these groups, and debates highlight not just scientific and economic issues but also moral questions about the relative 'purity' of chemically assisted food production versus 'clean' organic products, and the propriety of agricultural methods that may compromise longer term ecological health.

Engagements with water in the Stour Valley therefore contribute both materially and imaginatively to people's health and wellbeing in highly direct and powerful ways. Permeated by water, the body is both kept 'in order' and linked to wider systems of orderly flows. Envisioning this order provides ways to think about emotional and mental health flows and balances, and to consider larger scale material and non-material processes, ranging from the health represented by agricultural production or biodiversity, to more abstract spiritual issues.

## Antipodean water

The notion of water as essential to health and wellbeing at all levels is equally visible, albeit in an antipodean cultural form, in the beliefs of indigenous Australians living along the Mitchell River in Far North Queensland, with whom I conducted extensive ethnographic fieldwork at various times between 1991 and 2011. This research was conducted primarily with the Kunjen language speakers, but also with members of the Kokobera and Yir Yoront language groups who are now based in Kowanyama, near the river estuary (Figure 2.4).

*Figure 2.4* Kunjen youth at Maggie's Well waterhole, Cape York, Queensland.
Source: Author.

Until the early 1900s, the multiple language groups benefiting from the resource-rich wetlands and savannahs of Cape York maintained traditional hunter-gatherer lifeways. Their land and water management was low key: they used fire systematically to clear scrub and encourage 'green pick' (fresh new grass) for game, and built small traps and weirs to provide a plentiful supply of fish. With careful population control, and a vast lexicon of knowledge enabling them to make use of diverse resources, they were able to sustain this way of life for many thousands of years.

The diaries of early explorers offer some insights into the health of indigenous Australians prior to European settlement. Although in 1688 William Dampier suggested that 'the inhabitants of this continent are the miserablist people in the world', his objections were mostly about their lack of material culture and different appearance:

> No houses, and skin garments, sheep, poultry and fruits of the earth, ostrich eggs, &c. as the Hodmadods [Hottentots] have ... They have great heads, round foreheads and great brows ... They are long-visaged and of very unpleasing aspect, having no one graceful feature in their face.
>
> (Flannery, 1998: 27)

But while other visitors also expressed their disapproval of indigenous people's relative 'nakedness' and were similarly fearful of what they saw as a 'fierce' appearance, they also noted levels of physical fitness suggesting that Aboriginal communities' varied, fresh and unrefined diet, constant exercise and relative social stability appeared – generally – to support good health. Thus, in 1770, Joseph Banks noted that in the Aboriginal community with which he interacted:

> Their eyes were in many lively and their teeth even and good; of them they had complete sets ... They were all of them clean limned, active and nimble.
>
> (Flannery, 1998: 50)

However, in the late 1800s and early 1900s, European settlers appropriated the land in Cape York[3] and forced Aboriginal people either to provide free labour on the newly established cattle stations, or to seek refuge under the authority of the Christian missions established along the coast of the Gulf of Carpentaria. Along with the violence visited upon the indigenous communities, which reached genocidal levels (Reynolds, 1987), the arrival of European infectious diseases also had major impacts on Aboriginal health (Dowling, 1997), as did the trauma of dislocation and dispossession (Dodson, 2010).

Because of their deep affective attachment to their 'country', many groups chose to remain on their homelands, living double lives as stock workers and servants for the cattle stations,[4] while also continuing to hunt and gather, and to maintain their religious beliefs and practices. In Cape York, some did not move to the missions until the introduction of award wages in the mid-1960s.[5] It was thus possible, until relatively recently, to conduct ethnographic research with elders who had grown up in accord with longstanding indigenous beliefs and customs. The account below is drawn from my ethnographic research in this area, and from the Kunjen and other language group elders' generous efforts to teach me about their culture.

In a traditional Aboriginal cosmos, water is as fundamental to health as it is possible to imagine. It is the great generative power that, manifested in the form of the Rainbow Serpent (or Rainbow), is the wellspring for all life. In the creative era commonly known as the Dreamtime, the Rainbow Serpent spewed forth all of the totemic ancestral beings who, through their actions as hunter-gatherers, provided an exemplar for such lifeways, formed the landscape and its particular features, and made all of its living kinds. Having done so, they 'sat down' into the land where they remained as sentient ancestral forces, thus animating the entire landscape as an eternal living partner to its human inhabitants. In multiple forms – as rocks, trees and powerful water beings – the ancestral beings are believed to generate resources and the Law, thus meeting all of the physical, social, moral and spiritual needs of indigenous communities. Their continued presence reveals the reality that the Dreamtime is a spatial rather temporal concept, in essence representing an invisible non-material domain that coexists with the visible, material world. Thus, in discussing this era (more usually called the Story Time in northern Australia), the elders are more inclined to say 'where' rather than 'when'.

As a Kunjen elder explained, 'we all come from the Rainbow' (Lefty Yam, 1992). Held in the waters of the land, the Rainbow Serpent generates human spirits, which 'jump up' from water places to animate the foetus in a woman's body, providing some antipodean resonance with the previously mentioned linkage between the womb and water. Locating each individual as 'coming from' a specific totemic home[6] within the clan estate, this generative leap situates them not only within a network of kin, but also carries with it a share in collective clan rights to land and resources, ensuring their social and economic health and wellbeing. Life cycles are conceptualised hydrologically: people spring from a water place, and when they die, having circled through life to become 'closer to the ancestors', their spirit must be ritually returned to this home to be reabsorbed into the pool of ancestral power. From there, it will be reincarnated: thus, in kinship terminology, grandchildren may be described as 'little grandfather' or 'little grandmother', and will often inherit a grandparent's name when that person dies.

In effect, the Rainbow Serpent is a classic hydrotheological cycle, conflating the material flows of water at local and larger scales with the movements of the spirit, and of persons between material and non-material domains of existence, and between 'under' and 'over' worlds. In Aboriginal (and one might say Deleuzian[7]) terms, 'becoming' a person entails emerging from the contiguous Dreamtime and 'becoming visible' or 'becoming material' (Morton, 1987), alongside all of the other creative outputs from the Rainbow.

The Rainbow Serpent is also seen as the source of the Ancestral Law. This term encompasses all Aboriginal cultural knowledge, passed on through songs, stories, art and performance: the close-grained understanding of local topography, flora and fauna; the skills required for successful use and management of resources; the rules governing every area of activity; and the deeper levels of secret sacred knowledge to which people are initiated at critical life stages in order to become elders and provide social and moral leadership for communities.

The Rainbow Serpent may therefore be described as an ordering principle, through which all aspects of life maintain a 'proper way' of being. There are some direct relationships with physical health. For example, while Western medical notions of irrigating the body to ensure good health may only have traction with younger generations, the lifelong ingestion of homeland water and food is traditionally seen as fundamental to good health, these being the materials that 'grow up' the person so that they are properly (i.e. substantially) composed of their land and its waters.

However, as in the previous example, concepts of health shift readily from micro to macro levels. Water enables people to conceptualise flows between the individual and the wider human and non-human environment. An illustration of this interconnection is provided by ideas about how personhood is formed: this is seen as occurring via the infant's open fontanel, through which the flow of knowledge/water is seen to carry knowledge and consciousness into the person. Thus, individual knowledge is linked with the collective ancestral knowledge contained in sacred water places (Morphy and Morphy, 2014). This connection coheres with a belief that at the end of life, individual consciousness will flow back into the waters of the land, in a process of 'forgetting' that accompanies the person's shift back to the fluidity of the formless non-material domain (Frances Morphy, pers. comm.).

Living with sacred water places is also important in terms of phenomenological engagement. Aboriginal people do not discuss this via ideas about being 'mesmerised' or 'hypnotised' by the shimmering water, but there is a clear implication, demonstrated in highly respectful rituals, that such places elicit powerful affective responses. As well as outlining the spiritual and social movements of emerging from the Rainbow, ancestral stories often describe transformations caused by engagement with water places: changes in physical wellbeing or emotional state, and the acquisition of illuminating insights. For example, a particularly important ritual known as 'passing through the Rainbow' involves immersion in a powerful sacred water place, and confers deep spiritual knowledge and authority.

Moral wellbeing in these and all matters is aided by the ongoing presence of the ancestral beings who, as well as watching and nurturing, are also believed to punish transgressions and to withhold resources from strangers. Those who fail to respect sacred places risk being swallowed by the Rainbow Serpent or having other misfortunes visited upon them. When an elder dies, the waterhole associated with them might be 'sorry' and dry up for some time. Similarly, if sacred places are harmed, the related clan members may fall ill, or die. Even the presence of strangers can be painful: as a Kunjen elder explained, he felt 'too many footprints on me' caused by the incursion of tourists into his homelands (Nelson Brumby, 1992; Figure 2.5).[8] Thus, in both spiritual and physical terms, human and environmental health are closely interdependent.

A vision of the orderly flows of water that compose the person and their connections with kin and country is readily transposed into a broader collective relationship with the wider environment. Traditionally, the Law requires careful land and water management, aimed at maintaining healthy and productive ecosystems in ways that have been demonstrably sustainable for millennia. Primarily this entailed avoiding major population expansion and overuse of resources, as well as the fire management noted above. A key aspect of maintaining social and environmental health, however, is that concerned with flows of water.

Aboriginal communities across Australia share a keen sense of the importance of enabling and not disrupting or impeding the 'proper' water flows essential to environmental (and thus human) health. Customary indigenous water management was very low key, involving small fish traps and weirs and – in some areas – small channels to irrigate favoured plant species. It is therefore unsurprising that mining, which entails digging traumatically into the ancestral domain and disturbing (and sometimes releasing poisons into) water courses, has often generated anguished protests from indigenous communities. An example is the famous bark petition sent from Yirrkala to the Australian government in 1963, to appeal against the granting of mining rights to Nabalco in Arnhem Land (Commonwealth of Australia, 1963). The petition not only highlighted the threat to Aboriginal livelihood posed by

*Figure 2.5* Nelson Brumby, Kunjen elder.

the bauxite mine, but noted that this was inextricably linked with 'the places sacred to the Yirrkala people'. In so doing, it fuelled the major debates about land and water rights that followed. These have included many expressions of concern by Aboriginal communities about the effects of mining and cattle farming on local rivers and waterholes and, more recently, anxieties about the building of increasingly large dams and diversions of water to supply urban areas and provide crop irrigation. Merlan therefore reports how, in northern Australia, disruptions to the local landscape led to concerns that this would 'kill' its resident rainbow serpents (1998, see also Strang, 2004 [2001]).

At the heart of such concerns is a view that such impositions constitute a traumatic disruption of the ancestral domain, and of human and non-human health and wellbeing. On one level, this focuses on material issues, such as the effects of leaching of cyanide from old mine tailings into water courses and aquifers. There is also the impact of thousands of cattle station and farm bores drilling into and reducing the levels of the Artesian Basin, so that indigenous communities (often located around its margins in the north) find that the water quality in local wells is compromised. But, more fundamentally, such concerns conform to the principles of health and order held within Aboriginal Law, and how this conceptualises 'proper way' – i.e. orderly – flows of water, resources, people and spiritual being. The colonial dispossession of the traditional owners is seen as intrinsically disruptive to their abilities to ensure, through practical and ritual engagement with the ancestral land and waterscape, that social, environmental and spiritual health are maintained.

Following the bark petition protesting about Nabalco's bauxite mine, it took another 30 years of campaigning before the Australian government re-linquished the convenient calumny of *terra nullius* and acknowledged, with the Native Title Act of 1993, that Aboriginal people had their own forms of land ownership prior to European settlement of the continent (Williams 1986). While some subsequent land claims have been successful, a more common outcome of this shift has been Indigenous Land Use Agreements with non-Aboriginal landowners, and the greater involvement of Aboriginal people in land and water management. Such involvement continues to be guided by a uniquely indigenous vision of health in relation to water.

## Conclusion

The two cultural contexts through which this chapter has explored water and health are unique, and in many ways could not be more antipodean to each other. But there are some interesting commonalities. In both, it is readily evident that cognitive engagement with the materialities of water involves using it to think with, and to conceptualise notions of health as a matter of orderly systemic flows within the body, and between the body and the wider material environment. Both make seamless scheme transfers that allow similarly systematic views of the flows of health and wellbeing of larger systems that, like the body, require flows to be timely and sufficient, and – above all – unpolluted by the wrong kinds of matter.

Both world views employ images of water to envisage flows of wealth and resources, with some diverse valorisations of what constitutes healthy production and reproduction. In less tangible terms, water also appears in both as the medium through which it is possible to imagine the flows of knowledge – both enlightening and polluting – that may support or detract from social and environmental health.

There is also strong resonance in these communities' respective visions of water as a generative life-source: the essential substance from which all life is created, and which maintains health throughout all human and non-human flows of existence. There is some related coherence in ideas about the fluid substances that compose the self, and ensure that identity maintains wholeness and integrity. In Dorset, identity is seen in terms of blood relations, whereas in Cape York it is more a matter of being composed of the substance of the country itself, but in both cases it relies on concepts of substance and flow.

This draws attention to another area of commonality, which is the appreciation of the relationship between water and health in spiritual terms. Both world views 'think with' water flows to imagine the spirit and its movement between material and non-material domains. Both have complex baptismal and mortuary rituals to articulate the ways in which such movements compose orderly social and environmental relations. Further, in both cases, the spiritual and social health of communities is intimately connected by water to the health and wellbeing of local and wider ecosystems. The chapter's ethnographic comparison therefore provides insights into shared, as well as culturally specific, ideas about the meaning of water in relation to health, and how it enables conceptual flows across different domains.

## Notes

1 At that time, Wessex Water was owned by Enron. People also recalled the major error in the treatment of water by the geographically nearby South West Water Authority, which in 1988 had poisoned a number of people in Camelford, Cornwall (Morris, 2012).
2 By flooding the meadows and preventing the grass below from freezing, such water meadows extend the period during which it is possible to grow fodder for cattle.
3 Being geographically distant from the major areas of colonisation further south, Cape York was not settled by Europeans until a gold rush in the late 1800s encouraged the establishment of cattle stations to supply the influx of miners.
4 Both men and women worked on stock teams, and as servants, and some women were forced into concubinage.
5 The new law required cattle station owners to pay Aboriginal people for their labour – but as many cattlemen chose not to do so, this led to the eviction of the local clans from the stations.
6 In Kunjen, such places are called *Errk elampungk*, which composed of terms for 'place' (*errk*), 'eye' (*el*) and 'home' (*ampungk*), and may be translated as 'the home place of your image', underlining the shift from non-material to material being.

7 There is nice resonance here with the notions of 'becoming' articulated by Deleuze and Guattari (2004).
8 Nelson Brumby is now deceased. Aboriginal custom generally eschews the use of personal names for some time after an individual dies, but it is now many years since this occurred, and Nelson's name has been passed on – as is also customary – to a grandchild, and can now be used.

## References

Atran, S. (1990). *Cognitive Foundations of Natural History*. Cambridge, New York, Cambridge University Press.
Bachelard, G. (1983). *Water and Dreams: An Essay on the Imagination of Matter*. Dallas, TX, Pegasus Foundation.
BBC News. (2016). *Children 'Taken Ill After River Stour Swim' in Dorset*. Available at: http://www.bbc.co.uk/news/uk-england-dorset-37082507, last accessed August 15, 2016.
Bourdieu, P. (1977). *Outline of the Theory of Practice*. Cambridge, Cambridge University Press.
Chen, C., Macleod, J. and Neimanis, A. (eds) (2013). *Thinking with Water*. Montreal, McGill-Queens University Press.
Cohn, S. (1997). Being told what to eat: conversations in a diabetes day centre. In, P. Caplan (ed), *Food, Health and Identity*. London, New York, Routledge, 193–212.
Commonwealth of Australia, House of Representatives. (1963). *Report of the Select Committee on Grievances of Yirrkala Aborigines, Arnhem Land Reserve 1963*. Canberra, AGPS.
Corbin, A. (1994). *The Lure of the Sea: The Discovery of the Seaside in the Western World 1750–1840*. Cambridge, Polity Press.
Daily Echo. (2011). *Stour in Top Ten Most Improved Rivers*. Available at: http://www.bournemouthecho.co.uk/news/9221897.Stour_in_top_ten_most_improved_rivers, last accessed August 31, 2011.
Deleuze, G. and Guattari, F. (2004). *A Thousand Plateaus*. London, Continuum.
Dodson, M. (2010). The dispossession of indigenous people and its consequences. *Parity*, 23 (9). Available at: https://openresearchrepository.anu.edu.au/bitstream/1885/10682/1/Dodson_DispossessionIndigenous2010.pdf, last accessed July 27, 2011.
Dowling, P. (1997). A Great Deal of Sickness: Introduced Diseases among the Aboriginal People of Colonial Southeast Australia 1788–1900. *PhD Thesis*, Australian National University.
Edgeworth, M. (2014). On the agency of rivers. Commentary. *Archaeological Dialogues*, 21 (2), 159–162.
Féaux de la Croix, J. (2011). Moving metaphors we live by: Water and flow in the social sciences and around hydroelectric dams in Kyrgyzstan. *Central Asian Survey*, 30 (3–4), 487–502.
Flannery, T. (1998). *The Explorers*. London, Phoenix.
Foley, R. (2010). *Healing Waters: Therapeutic Landscapes in Historic and Contemporary Ireland*. Farnham, Ashgate.
Foley, R. (2011). Performing health in place: The holy well as a therapeutic assemblage. *Health and Place*, 17, 470–479.

Foley, R. (2015). Swimming in Ireland: Immersions in therapeutic blue space. *Health and Place,* 35, 218–225.

Illich, I. (1986). *H₂O and the Waters of Forgetfulness.* London, New York, Marion Boyars.

Ingold, T. (2010). Footprints Through the Weather-World: Walking, breathing, knowing. *Journal of the Royal Anthropological Institute,* Special Issue 2010. S121-S139.

Kaplan, S. (1995). The restorative benefits of nature: Toward an integrative framework. *Journal of Environmental Psychology,* 15, 169–182.

Kelly, M. (2015). *Quartz and Feldspar: A History of Modern Dartmoor.* London, Jonathan Cape.

Lakoff, G. and Johnson, M. (1980). *Metaphors We Live By,* Chicago, London, University of Chicago Press.

McMenamin, D. and McMenamin, M. (1994). *Hypersea.* New York, Colombia Press.

Merlan, F. (1998). *Caging the Rainbow: Places, Politics and Aborigines in a North Australian Town.* Honolulu, University of Hawai'i Press.

Morphy, F. and Morphy, H. (2014). We Think Through Our *Marwat* (paintbrush): Reflections on the Yolngu Location of Thought and Knowledge. *The Robert Layton Lecture,* October 8th, 2014. Durham University.

Morris, S. (2012). Camelford case coroner accuses water authority of gambling with 20,000 lives. *The Guardian.* Available at: https://www.theguardian.com/society/2012/mar/14/camelford-case-coroner-water-poisoning, last accessed March 14, 2012.

Morton, J. (1987). The effectiveness of totemism: Increase rituals and resource control in Central Australia. *Man,* 22, 453–474.

Ohly, H., White, M., Wheeler, B., Bethel, A., Ukoumunne, O., Nikolaou, V. and Garside, R. (2016). European Centre for Environment and Human Health, University of Exeter Medical School, Truro Campus, and Knowledge Spa, Royal Cornwall Hospital, Truro, Cornwall, United Kingdom view further author information attention restoration theory: A systematic review of the attention restoration potential of exposure to natural environments. *Journal of Toxicology and Environmental Health, Critical Reviews,* 19 (7), 305–343.

Reynolds, H. (1987). *The Law of the Land.* Ringwood, Penguin.

Sassoon, S. (1909). *Everyone Sang.* The Poetry Foundation. Available at: https://www.poetryfoundation.org/poems/57253/everyone-sang, last accessed December 1, 2018.

Suedfeld, P. Ballard, E. and Murphy, M. (1983). Water immersion and flotation: From stress experiment to stress treatment. *Journal of Environmental Psychology,* 3, 147–155.

Strang, V. (2004) [2001]. Poisoning the rainbow: Cosmology and pollution in Cape York. In, A. Rumsey and Weiner, J. (eds), *Mining and Indigenous Lifeworlds in Australia and Papua New Guinea.* Wantage, Sean Kingston Publishing, 208–225.

Strang, V. (2004). *The Meaning of Water.* Oxford, New York, Berg.

Strang, V. (2014). Fluid consistencies: Meaning and materiality in human engagements with water. *Archaeological Dialogues,* 21 (2), 133–150.

Williams, N. (1986). *The Yolngu and Their Land: A System of Land Tenure and the Fight for Its Recognition.* Canberra, Australian Institute of Aboriginal Studies.

# 3 Keeping leisure in mind

## The intervening role of leisure in the blue space – health nexus

*Sean Gammon and David Jarratt*

### Introduction

Blue space is a term coined by environmental and social psychologists to describe predominantly aquatic environments such as riversides, seashores, lakes and oceans. Recently, an increasingly significant body of literature has associated these environments with wellness-related benefits such as a feeling of restoration (Foley and Kistemann, 2015). This chapter acknowledges advances in this area, considers the nature of this blue space – health nexus and focuses on the influence of leisure upon it. We contend that leisure and associated states of mind play a vital role in the relationship between health/wellness and blue space.

The focus here is the coast rather than all aquatic environments; this is the most obvious example of blue space that is clearly also a leisure zone. More leisure tourism takes place on or near the coast than any other type of environment; for example, 51% of bed capacity in European hotels is concentrated in coastal areas (European Commission, 2017). In 2015, seaside locations accounted for 39% of British holiday nights and there were 14.91 million trips to British beaches, which were associated with £3.84 billion of visitor spending (Visit Britain, 2017). The seaside resort of Blackpool (UK) saw visitor numbers increase to 18 million in 2016 (Marketing Lancashire, 2017), a trend which was mirrored nationally.

Before moving onto the significant role of leisure, it is useful to provide a brief history of blue leisure space, with specific reference to wellness. The emphasis here is on the British seaside and the resorts that lie at the heart of the UK's coastal visitor economy. These were the first places to offer leisure tourism on an industrial scale; they were the facilitators of exposure to blue space for millions of people and continue to serve this purpose. As this chapter progresses, it should be apparent that observations regarding the function of leisure should, in principle, apply equally to other industrialised societies even if these are differentiated by seasonal differences in access to and engagement with the coast.

## The invention of the seaside: leisure and blue space
## for the masses

The meanings and interpretations of the seaside have varied and changed through history (Andrews and Kearns, 2005; Gillis, 2012). For example, the industrial and post-industrial British seaside resort has been viewed as a site of the Carnivalesque (Shields, 1991) and nostalgia (Jarratt and Gammon, 2016; Walton, 2000). Among the best-known and oldest associations with the British seaside resort is that of wellbeing. It became the driving force behind mass coastal leisure tourism in the industrial age. Popular seaside resorts, such as Scarborough, were direct descendants of the inland spa (see Gesler, 1998). The development of these resorts is now well documented by historians (see Brodie, 2018; Walton, 2000). A full review of this development lies outside the scope of this chapter, which instead considers the motivation of visitors to engage with these resorts.

The seaside pioneers, the British upper and middle classes, were motivated to visit the seaside partly because they felt increasingly isolated from the rhythms of nature, as industrialisation gained momentum (Corbin, 1994; Lenček and Bosker, 1998). In the age of early industrialisation, people returned to the sea in a search for something that was missing from new towns and cities – *wilderness*, which is what the awe-inspiring sea came to represent to Western urbanites (Corbin, 1994; Gillis, 2012). The medical profession in the 18th century seemed to underline the need for rest and treatments to alleviate anxiety and ailments through this powerful tonic. Corbin (1994: 62) writes, 'The sea was expected to cure the evils of urban civilization and correct the ill effects of easy living'. The seaside was 'invented' through the harnessing of the 'healing power' of this environment. The medicalised seaside was to play a key role in the industrial society that had supported it as Perkin (1970: 224) explains,

> The seaside resorts as a whole had grown up to meet a need of industrial society as urgent as the need for Lancashire cottons or Birmingham hardware. They were, indeed, industrial towns with a specialised product, recreation and recuperation, made necessary by the growth of other specialised towns and the concentration of most of the population in an urban environment from which periodical escape was important for mental and physical health.
>
> (1970: 224)

Seaside leisure tourism grew rapidly in late 19th-century Britain as well as across much of Europe, North America and Australasia (Walton, 2000). The touristic infrastructure and seaside entertainments developed to cater for growing numbers of visitors. Industrial workers who had more disposable time, money and affordable transportation options than their rural counterparts were a targeted clientele. Yet their holidays had were often rooted in agricultural holidays of old. Arguably older carnivalesque elements were carried over from these older traditions to the seaside resorts with their new entertainments (Hyman and Malbert, 2000; Shields, 1991). Nevertheless, health and wellness were still a significant element of what resorts offered (Walton, 2000). In 1930s Lancashire, the most significant motivations for Boltonians to visit the resort of Blackpool were: holiday (leisure) pastimes, rest and relaxation, health-seeking, romance, a change from Bolton and 'nature'. So, while pastimes and pleasure-seeking were of crucial importance, rest, relaxation, health and nature all take their place as important motivators for interwar seaside leisure (Cross, 1990: 43).

Post-war Britain saw a seaside boom; in developed countries visitors were now more likely to top up their suntan instead taking the waters or the 'ozone' (Collins and Kearns, 2007; Walton, 2000). From the 1960s onwards, the middle classes could afford Mediterranean package holidays. Increasingly the seaside holiday was therefore not to be restricted to one's home country as we entered an age of cheaper air travel. In addition, stories of water pollution around the British shoreline surfaced in the second half of the 20th century. The traditional British seaside resort now had much more competition and faced a challenging end to the 20th century. The 21st century has marked a period of renewed interest in these resorts and record-breaking growth in the UK's coastal visitor economy, as mentioned earlier. Internationally, the growth of coastal visitor numbers and development shows no sign abating; indeed, it provides a significant challenge in terms of sustainability. Wellness continues to be an important element of the coastal visitor's experience as we discuss in the next section. However, before doing that, the connection between wellness/restoration and the coast needs to be explored.

### The contemporary seaside – a restorative 'blue space'?

> Nature's most potent antidepressant, the beach moves us with a power of a drug, the rhythm of its tides and shifting margins re-orientating our sense of space and time.
>
> (Lenček and Bosker, 1998: xix)

Gallagher (2007: 228) suggests that society has been looking inwards and 'living in the space between our ears' and that we have 'forgotten what our forbearers knew and our scientists are rediscovering'. In other words, we have become disconnected from the natural environment and scientists and psychologists now observe how exposure (or reconnection) to natural environments offers wellness-related benefits. Kaplan and Kaplan (1989) suggested that restorative natural environments offer an optimal experience which can ease mental fatigue; Attention restoration theory argues that such environments can nourish attention and replenish depleted energy. Of course, we should not assume that all people(s) perceive natural environments in an equivalent way; indigenous peoples often feel a more immediate connection to nature and place (see Smith, 2008).

In recent years, research has focused more specifically on blue spaces and associated wellness benefits (see Foley and Kistemann, 2015; Gascon et al., 2017). Researchers associated with The European Centre for Environment and Human Health (www.ecehh.org) have linked positive feelings (for example, calm, refreshment and enjoyment) with exposure to the coastal environment. White et al. (2010) acknowledge that the reasons for the restorative nature of natural environments and blue space are unclear, but suggest three potential explanations:

a    Visual properties, for example, the reflection of light in interesting and restorative ways.
b    These environments are associated with restorative sounds, for example, breaking waves.
c    The possibility (real or imagined) or immersing oneself in water and the associated drop in stress levels.

The nature of the seaside experience and the related state of mind of the visitor hold the key to this puzzle. The 'emptiness' of the coast allows for a slowness; strolling, 'pottering' and generally slowing down are of vital importance to the seaside experience today (Baerenholdt et al., 2004: 32). Holidays can offer catharsis and escape from responsibilities and everyday constraints of time. The liminality of the holidaymakers' beach marks an evident break from everyday behaviour. For instance, it's a place and time where various states of undress are not just tolerated but expected and one of the few places where adults can have unstructured childlike fun (Ryan, 2010; Shields, 1991). The importance of time to leisure and travel has been discussed: from Tuan (1977: 122) who observes that timelessness has been associated with some holiday destinations to taking time to savour leisure experiences (Kurtz and Simmons, 2015).

According to Gillis (2012), a sociocultural view of the seemingly timeless sea and empty coast emerged in the late 19th century. As Europeans and North Americans became less physically connected to the coast they became closer to them mentally and imaginatively. 'In its new role as wilderness, the sea offered an escape from time, from history itself' (Gillis, 2012: 141). Gillis also observes that those who derive their living from the sea are unlikely to hold such abstract and landlubber-like views. The timeless, eternal and romantic sea can be considered 'modern' – constructed by those who are alienated from it. This is in stark contrast to our ancestors who were of the edge and did not make as sharp a distinction between land and water; to use a 17th-century term, they saw the world as 'terraqueous' (Gillis, 2012: 99). Finding blue space may be a tonic for contemporary people but the drive to do so is also a symptom of industrialisation and urbanisation, for 'the Industrial Revolution drew the West indoors' (Gallagher, 2007: 13).

There are growing concerns that children are spending too much time inside and in front of screens and not enough time outside; among the most commonly cited statistics in this area is that the average American now spends over 90% of their time inside a building or vehicle (Gallagher, 2007).

This modern-day reconnection, as described by Gillis, has been considered by Jarratt (2015a), Jarratt and Gammon (2016) and Jarratt and Sharpley (2017) who researched the experience of older visitors to the traditional British seaside. A distinctive sense of place (coined 'seasideness' by Jarratt, 2015a) manifested itself through positive and uplifting experiences of the multisensory seaside environment; interviews reveal a belief that the seaside was a 'tonic'. The open vistas, fresh air, the smell of the sea and sound of crashing waves were consistently referred to and described in restorative terms. Furthermore, this distinctive environment was seen to offer a temporary escape from the pressures of contemporary life; interviewees felt themselves relaxing, adjusting to natural rhythms and slowing down – these experiences were a key part of seasideness and visitor appeal. Furthermore, this research suggests that the natural seaside environment held qualities that encouraged contemplation. Interviewees considered their seaside to have a meaningful element that is best described as *spiritual* (Jarratt and Sharpley, 2017). The most common interpretation has been a secular spirituality which involved a (re)connection with the natural world through appreciating its beauty, complexity, scale and timelessness. Such perceptions of the sea were central to these potentially spiritual reactions (Jarratt, 2015b). The visitors saw the seaside experience as an opportunity to reconnect with something bigger or more significant than themselves (the natural world), while simultaneously confirming their place within it, and so it did indeed take on a spiritual significance. This research does not stand in isolation; while few other scholars have addressed the role of the seaside environment in the *touristic* sense of place, sense of place at the coast more generally has been considered (for example, Wylie, 2005). A particularly resonate study was conducted by Bell et al. (2015) into the therapeutic nature coastal experiences – in this study participants expressed strong and enduring connections to their local coastline.

For hundreds of years the seaside has been linked to holistic benefits for both body and mind. Most recently, this link has been explored by psychologists, but in terms of leisure tourism this appreciation of the coast can be traced back at least as far as the 17th century, but remains to this day. The seemingly 'timeless' seaside offered an opportunity to reconnect, tune into natural rhythms, slow down and contemplate in a place that contrasted with the frenetic pace of modernity (Jarratt and Gammon, 2016; Jarratt and Sharpley, 2017). According to Gillis (2012) and Jarratt (2015b), the most significant feature of the coast that enables all this is a blankness or emptiness – of human-made distractions, of history and, especially, of our notion of time. Ironically perhaps, this view of the sea is relatively modern and, of course, anthropomorphic. Gillis argues that since the industrial revolution we have turned our back and the sea emptied our coasts and we now seek to return to them. A key feature of this return, especially from a visitor perspective, has been, and still is, the perception of wellness. Of course, such perceptions may vary, but the aforementioned research by psychologists appears to confirm this connection. Whether it be socioculturally constructed or not, it is easy for us to conceive a place of slowness, leisure, reverie and introspection as something of a tonic.

## Leisure in blue space

As detailed above there is mounting evidence of the many potential health benefits gained by being in and around blue spaces. However, the overwhelming majority of studies tend to explore either the health-giving properties of the environment or the resulting impacts on the individual. What is missing is any consideration of the psychological state of those benefiting from such environs, before, during and after interaction. For, as in radio transmission, the quality of the communication is largely dependent upon the condition of the receiver. Since a significant proportion of visits to the coast are leisure-related, it would seem logical to explore how being in and/or at leisure may affect the extent of any given benefit. Consequently, we ask: to what extent does leisure positively enhance the potential health benefits from the environment? Before exploring this question further, it is first necessary to outline, from the many meanings and definitions that the term leisure attracts, which is pertinent to this proposition.

It is beyond the scope of this chapter to offer a comprehensive discussion on the numerous manifestations of leisure (see Best, 2010; Bull et al., 2003; Elkington and Gammon, 2014; Page and Connell, 2010; Torkildsen, 2005). Yet it is possible to indicate the salient approaches that have been taken to enhance our understanding of it. In simple terms, leisure can be explained from either temporal, objective or subjective viewpoints (Stockdale, 1985). Temporal approaches view leisure as those *activities* that take place during periods perceived as being free from obligation. The primary focus of studies that take this approach has been to determine the differences and potential inequalities in leisure opportunities between individuals and various social groups. It is important to note that an integral feature of the temporal component of leisure is that *leisure time* is not synonymous to *free time* – as *free time* can be both imposed and/or enforced, whereas leisure time is welcomed and valued.

The activities that take place during such leisure episodes occupy the interest of those who take a more objective approach to leisure. In this case, it is what people choose to do – or not do – during their leisure that is deemed important. The applications of this approach are many and varied, including the health consequences of active and passive leisure (Iso-Ahola, 1997), the social consequences of positive leisure episodes in adolescence (Mahoney and Stattin, 2000), the impact of leisure on the elderly (Silverstein and Parker, 2002), identity acquisition and affirmation in leisure (Haggard and William, 1992) and numerous studies that explore the choices and constraints of leisure between social groups (Critcher, 2006; Roberts, 1999). While there is little doubt that studies exploring the activities undertaken offer important insights into leisure's meaning and purpose, they also highlight the problems of delineating specific activities as leisure. Even though in most developed countries there is a recognised 'leisure industry', there is a distinct lack of agreement as to the leisure status of all the activities offered. As a result, leisure activities are only related to what people choose to do in their leisure, rather than being prescriptive pursuits that denote indisputable leisure episodes.

To offer some continuity and agreement as to what constitutes leisure, psychologically driven studies have argued that, irrespective of what individuals choose to do with their leisure, the one constant is the experience itself. In other words, leisure is a distinct and unique state of mind that is common to all those who experience it. This leisure state has been identified as comprising perceived freedom blended with intrinsic and extrinsic motivation (Neulinger, 1976) and has a number of important implications. First, it releases leisure from the dependence on work; no longer considered as little more than residual time left over after work or other obligations, leisure can now be experienced at any time or place. Second, it can now be valued as something desirable in and of itself: 'It has become a state of mind brought about by a feeling of freedom and a feeling of doing something worthwhile, both highly cherished values in our society' (Neulinger, 1976: 17). Third, it brings attention to the numerous therapeutic features that the experience of leisure can have upon an individual (Mannell and Kleiber, 1997) – a factor that will be explored further in the following section.

## Leisure and health

The notion that engaging in leisure has numerous health benefits has been discussed extensively in the literature (Caldwell, 2005; Driver et al., 1991; Haworth, 1997; Newman et al., 2014; Page and Connell, 2010). However, the overwhelming majority of studies have tended to explore the health-enhancing benefits of the chosen activities – rather than the physiological and psychological rewards of simply being at leisure. Nevertheless, there remains a significant body of research that has explored the numerous benefits to a person's health and wellbeing by entering the leisure state (Caldwell, 2005; Lloyd and Auld, 2002; Sonnentag, 2012; Watkins and Bond, 2007). For example, Lloyd and Auld (2002) found there to be a direct correlation between leisure satisfaction and life satisfaction. Furthermore, freely entering states that are primarily intrinsic in nature cultivate feelings of competence, life meaning and purpose, relaxation and focus. Not only do such experiences positively affect subjective wellbeing, but they also act as a buffer against negative life events through distraction and creating optimism about the future. This is not to suggest that all individuals will accrue similar benefits from entering leisure, as this will depend on a number of factors – one of which is value (Neulinger, 1981). It is unlikely there will be any significant health benefit of leisure when there is little or no intrinsic value present. If leisure is sought only for its extrinsic value then it just becomes nothing more than a means to an end. Therefore, the extent to which any chosen activity in leisure is rewarding and enjoyable is dependent upon whether it is appreciated and deemed a justifiable use of time. As Iso-Ahola (1997: 132) notes: 'The fact that an individual acknowledges, values and engages in leisure for its own sake, for its inherent characteristics, is one way in which leisure contributes to health'. If leisure is not valued then the extent to which it positively impacts upon health is significantly reduced (Neulinger, 1981).

A further outcome of entering the leisure state relates to agency: the ability for people to immerse themselves in whatever they have chosen to engage in. Csikszentmihalyi's (1988, 1992) work on flow is often used as an example of individuals commonly immersed in high investment activities to such a degree that little else matters – except for what is in front of them. Csikszentmihalyi (1988: 170) refers to the resulting benefit of these types of experiences as psychic negentropy – a 'state we usually describe by such terms as "joy," "happiness," "satisfaction," "clarity" or "sense of achievement"'. Obviously, powerful outcomes like these have beneficial effects on health and wellbeing. However, it is important to note that while flow-like experiences are not the sole preserve of leisure episodes, the more powerful flow experiences are more likely to take place in activities freely chosen that generate intrinsic interest to the recipient.

Savouring is another form of focus, which in simple terms refers to those moments when someone can '...attend to, appreciate, and enhance the positive experiences in their lives' (Bryant and Veroff, 2007: 2). Therefore, when there is little or no distraction and time to soak in and cherish a meaningful life episode, then savouring takes place. It can be triggered by external stimuli like an awe-inspiring landscape (Gammon and Elkington, 2018), or can emanate internally through an appreciation of a particularly pleasant mindset. The outcome of these types of events are many and varied, though primarily revolve around adding clarity to any given situation and potentially contributing to individuals' personal growth:

> ... focused attending gives positive stimuli greater emotional power, gives people greater access to their feelings in response to these stimuli, and establishes a more reliable way to recall and relive a positive experience later. We suggest that focussed attending makes a positive experience more distinctive, more vivid, and more easily and fully savored.
>
> (Bryant and Veroff, 2007: 69)

Similarly to flow, savouring can take place in moments outside leisure but is more likely to take place when a sense of freedom and intrinsic interest is evident. Furthermore, savouring and leisure have been identified as being closely related, each positively influencing the other (Kurtz and Simmons, 2015).

In sum, leisure is beneficial to health through the activities chosen, coupled with the inherent intrinsic properties that it produces. Of course, this is not to suggest that all choices in leisure are beneficial to health, as clearly long-term inactivity or the consumption of large amounts of alcohol, for instance, can have a deleterious effect on health and wellbeing. Moreover, such deleterious behaviour has directly been linked to some seaside resorts (Agarwal and Brunt, 2005). However, in these cases, the harmful consequences of the environment stem from the influences associated with the in-land aspects of a place, rather than the natural vistas that it offers. Yet, such harmful choices should not distract from the innate qualities of entering and valuing the leisure state, coupled with self-determined visits to coastal regions.

## Leisure, health and blue space

Choosing to visit blue (and/or green) spaces in leisure and the consequent health benefits of such choices have been recognised and acknowledged for centuries. Yet the overwhelming focus has been in exploring the health-giving properties of the environment, rather than the psychological mindsets of those benefiting from the surroundings. As Volker and Kistemann (2011: 458) affirm, 'Emotional and experiential responses to blue space have not yet been adequately recognised'. Equally, neither has the necessary mind states before entering blue spaces been fully acknowledged. As a result, the aim of this section is to make the case that to enter and engage with blue spaces when at leisure amplifies the benefits that such spaces emit. We hypothesise that leisure is a potential intervening variable – simply put, as BS + L $\land$ H. This proposition is based on the idea that individuals are more open and more sensitive to the health-giving properties of blue spaces when there is time to focus and savour the moment. In addition, a duality of value takes place – where the value and appreciation of experiencing leisure combines with the value and appreciation of the environment. The resulting mind state analogously equates to a widening of an aperture that (see Figure 3.1) enables the individual to be more receptive to the health-inducing properties of the environment.

When the mind is not distracted, there is time and motive to savour and soak in the environment, which may create a reciprocal benefit to both the experience of leisure and the appreciation of the surroundings. Furthermore, the feelings of wellbeing and mood improvement continue well after the interaction has taken place (Caltabiano, 1995), a condition Foley (2017) refers to as 'accretive practice'. As intimated above, there are still health-related advantages from working and living in and around blues spaces, but there

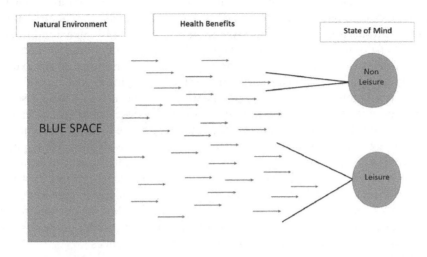

*Figure 3.1* The leisure health receptor model.

is a reduction in these benefits, primarily due to increased distraction and complacency towards the environment. The notion that all blue spaces endow intrinsic therapeutic benefits has come into question (Williams, 2007); the fact remains that, for many people, they offer sanctuary from the many pressures that modern life engenders. It has been suggested that the sense of wellbeing resulting from coastal areas arises through what Bell et al. (2015) refer to as four overlapping therapeutic experience dimensions: immersive, symbolic, social and achieving experiences. There is little doubt that such life-affirming experiences impact positively upon the wellbeing of the recipients, yet it is unlikely that such experiences would be achievable without first entering a leisure mind state.

## Conclusion

The many health-related benefits that can accrue from being in around blue spaces continue to occupy researchers across the globe. In simple terms, these benefits are derived from direct and indirect factors such as the positive physical and psychological consequences of being in such environments, or the symbolic and social experiences linked to wellbeing that blue spaces enable. However, there is good evidence to suggest that these positive outcomes do not occur osmotically; that to take full advantage of the varied benefits to health, individuals should first achieve the appropriate mind state. We have argued that the leisure state of mind can nurture, amplify and reify the health-giving properties of blue spaces and that leisure is the intervening variable that helps enable and accentuate the many positive experiences that are on offer.

The likely reason as to why the leisure state of mind should have such an affect is due, in part, to its own much-researched benefits to wellbeing. Blue spaces and leisure enjoy a symbiotic relationship, whereby each component benefits from the other. The positive health outcomes of choosing *to leisure* in natural environments are well documented, but the notion that the extent of these benefits is as much to do with the accompanying leisure state as it is to the environment has been largely neglected. What is more, little discussion has taken place on how the environment reinforces and enhances the leisure state. For example, experiences of freedom are integral elements of both being at the seaside and at leisure. But, there are also specific aspects of entering the leisure state that helps those visiting blue spaces to be more receptive to its potential benefits. To be at leisure at least minimises if not removes life's unwanted distractions. It allows us to be more open and welcoming to the environment. We are more likely to value, savour and focus on what is important and meaningful to us, while allowing ourselves the time to reflect and enjoy such transformational experiences.

While interest in the therapeutic nature of many natural environments continues to grow, the importance of the concomitant mind state has generated less interest. Consequently, future studies exploring the health benefits of blue spaces should not only explore the affects that the environment has upon the individual, but also to be cognisant of the accompanying mind state that occurs during such episodes. Of course, the experiences encountered when visiting coastal regions remain important but so too is the mind state when entering them. To gain a deeper and more nuanced understanding of the nature and processes involved in blue space-related heath, it is vital to explore why some individuals accrue deeper feelings of wellbeing than others. Future research might usefully consider how respondents feel while encountering blue spaces – and so to keep always leisure in mind.

## References

Agarwal, S. and Brunt, P. (2005). Social exclusion and crime in English seaside resorts: Implications for resort restructuring. *Tourism, Culture and Communication,* 16 (1), 19–35.

Andrews, G.J. and Kearns, R.A. (2005). Everyday health histories and the making of place: The case of an English coastal town. *Social Science and Medicine,* 60, 2697–2713.

Baerenholdt, J., Haldrup, M., Larsen, J. and Urry, J. (2004). *Performing Tourist Places.* Aldershot, Ashgate.

Bell, S.L., Phoenix, C., Lovell, R. and Wheller, B.W. (2015). Seeking everyday wellbeing: The coast as a therapeutic landscape. *Social Science and Medicine,* 142, 56–67.

Best, S. (2010). *Leisure Studies: Themes and Perspectives.* London, Sage.

Brodie, A. (2018). *The Sea Front.* Swindon, English Heritage Books.

Bryant, F.B. and Veroff, J. (2007). *Savoring: A New Model of Positive Experience.* Mahwah, NJ, Lawrence Erlbaum Associates.

Bull, C., Hoose, J. and Weed, M. (2003). *An Introduction to Leisure Studies.* Essex, Prentice Hall.

Caldwell, L. (2005). Leisure and health: Why is leisure therapeutic? *British Journal of Guidance & Counselling,* 33 (1), 7–26.

Caltabiano, M.L. (1995). Main and stress-moderating health benefits of leisure. *Loisir et Societé,* 18:92006 (1), 33–51.

Collins, D. and Kearns, R. (2007). Ambiguous landscapes: Sun, risk and recreation on New Zealand beaches. In, A. Williams (ed), *Therapeutic Landscapes.* Aldershot, Ashgate, 15–32.

Corbin, A. (1994). *The Lure of the Sea.* London, Blackwell.

Critcher, C. (2006). A touch of class. In, C. Rojek, Shaw, S. and Veal, A. (eds), *A Handbook of Leisure Studies.* Basingstoke, Palgrave, 271–287.

Cross, G. (1990). *Worktowners at Blackpool: Mass-Observation and Popular Leisure in the 1930s.* London, Routledge.

Csikszentmihalyi, M. (1988). Motivation and creativity: Toward a synthesis of structural and energistic approaches to cognition. *New Ideas in Psychology,* 6 (2), 159–176.

Csikszentmihalyi, M. (1992). *Flow: The Psychology of Happiness.* London, Rider.

Driver, B.L. et al. (eds) (1991). *Benefits of Leisure.* Champaign, IL, Sagamore.

Elkington, S. and Gammon, S. (eds) (2014). *Contemporary Perspectives in Leisure: Meanings Motives and Lifelong Learning.* London, Routledge.

European Commission. (2017). *Integrated Marine Policy – Coastal and Marine Tourism* [On-line PDF document]. Available at: https://ec.europa.eu/maritimeaffairs/policy/coastal_tourism_en, last accessed June 4, 2017.

Foley, R. (2017). Swimming as an accretive practice in healthy blue space. *Emotion, Space and Society,* 22, 43–51.

Foley, R. and Kistemann, T. (2015). Blue space geographies: Enabling health in place. *Health and Place,* 35, 157–165.

Gallagher, W. (2007). *The Power of Place.* New York, Harper Perennial.

Gammon, S. and Elkington, S. (2018). Landscape and leisure: A view worth seeing? In, P. Howard, Thompson, I., Waterton, E. and Atha, M. (eds), *The Routledge Companion to Landscape Studies.* London, Routledge, 366–375.

Gascon, M., Zijlema, W., Vert, C., White, M.P. and Nieuwenhuijsen, M.J. (2017). Outdoor blue spaces, human health and well-being: A systematic review of quantitative studies. *International Journal Hygiene and Environmental Health,* 220 (8), 1207–1221.

Gesler, W. (1998). Bath's reputation as a healing place. In, R. Kearns and Gesler, W. (eds), *Putting Health into Place: Landscape, Identity and Wellbeing.* New York, Syracuse University Press, 17–35.

Gillis, J. (2012). *The Human Shore: Seacoasts in History.* Chicago, Chicago University Press.

Haggard, L. and Williams, D. (1992). Identity affirmation through leisure activities: Leisure symbols of the self. *Journal of Leisure Research,* 24 (1), 1–18.

Haworth, J. (ed) (1997). *Work, Leisure and Well Being.* London, Routledge.

Hyman, T. and Malbert, R. (2000). *Carnivalesque.* London, Hayward Gallery Publishing.

Iso-Ahola, S. (1997). A psychological analysis of leisure and health. In, Haworth, J. (ed), *Work, Leisure and Well-Being.* London, Routledge, 131–144.

Jarratt, D. (2015a). Sense of place at a British coastal resort: Exploring 'seaside-ness' in Morecambe. *Tourism: An International Interdisciplinary Journal, 63* (3), 351–363.

Jarratt, D. (2015b). Seasideness: Sense of place at a seaside resort. In, S. Gammon and Elkington, S. (eds), *Landscapes of Leisure: Space, Place & Identities*. London, Palgrave Macmillan, 147–163.

Jarratt, D. and Gammon, S. (2016). 'We had the most wonderful times': Seaside nostalgia at a British resort. *Tourism Recreation Research, 41*(2), 123–133.

Jarratt, D. and Sharpley, R. (2017). Tourists at the seaside: Exploring the spiritual dimension. *Tourist Studies, 17* (4), 349–368.

Kaplan, R. and Kaplan, S. (1989). *The experience of nature.* New York, Cambridge University Press.

Kurtz, J.L. and Simmons, E. (2015). Savouring leisure spaces. In, S. Gammon and Elkington, S. (eds), *Landscapes of Leisure: Space, Place and Identities*. London, Palgrave Macmillan, 164–175.

Lenček, L. and Bosker, G. (1998). *The Beach: The History of Paradise on Earth.* New York, Viking.

Lloyd, K. and Auld, C. (2002). The role in determining quality of life: Issues of content and measurement. *Social Indicators Research, 57,* 43–71.

Mahoney, J. and Stattin, H. (2000). Leisure activities and adolescent antisocial behavior: The role of structure and social context. *Journal of Adolescence, 23* (2), 113–127.

Mannell, R. and Kleiber, D.A. (1997). *A Social Psychology of Leisure.* State College, PA, Venture Publishing Inc.

Marketing Lancashire. (2017). *STEAM – Tourism Economic Impacts* [On-line PDF document]. Available at: http://www.marketinglancashire.com/dbimgs/Marketing%20 Lancashire%20-%20STEAM%20Infographic%20-%202016%20-%20Final.pdf, last accessed June 4, 2017.

Neulinger, J. (1976). The need for and the implications of a psychological conception of leisure. *Ontario Psychologist, 8* (2), 13–20.

Neulinger, J. (1981). *The Psychology of Leisure.* Springfield, IL, Charles C. Thomas.

Newman, D., Tay, L. and Diener, E. (2014). Leisure and subjective well being: A model of psychological mechanisms as mediating factors. *Journal of Happiness Studies,* 15, 555–578.

Page, S. and Connell, J. (2010). *Leisure: An Introduction.* Essex, Prentice Hall.

Perkin, H. (1970). *The Age of the Railway.* London, Granada Publishing Limited.

Roberts, K. (1999). *Leisure in Contemporary Society.* Wallingford, CABI.

Ryan, C. (2010). Memories of the beach. In, C. Ryan (ed), *The Tourist Experience.* London, Thomson, 156–172.

Shields, R. (1991). *Places on the Margin: Alternative Geographies of Modernity.* London, Routledge.

Silverstein, M. and Parker, M. (2002). Leisure activities and quality of life among the oldest old is Sweden. *Research on Aging, 24* (5), 528–547.

Smith, A. (2008). A Māori sense of place? – Taranaki Waiata Tangi and feelings for place. *New Zealand Geographer, 60* (1), 12–17.

Sonnentag, S. (2012). Psychological detachment from work during leisure time: The benefits of mentally disengaging from work. *Current Directions in Psychological Science, 21* (2), 114–118.

Stockdale, J.E. (1985). *What is Leisure? An Empirical Analysis of the Concept of Leisure and the Role of Leisure in People's Lives*. London, Sports Council.

Torkildsen, G. (2005). *Leisure and Recreation Management*. London, Routledge.

Tuan, Y. (1977). *Space and Place*. Minneapolis, University of Minnesota Press.

Visit Britain. (2017). *The GB Tourist* [On-line PDF document]. Available at: https://www.visitbritain.org/sites/default/files/vb-corporate/Documents-Library/documents/England-documents/gb_tourist_report_2015.pdf, last accessed June 4, 2017.

Volker, S. and Kistemann, T. (2011). The impact of blue space on human health and well-being – Salutogenetic health effects of inland surface waters: A review. *International Journal of Hygiene and Environmental Health*, 214 (6), 449–460.

Walton, J.K. (2000). *The British Seaside: Holidays and Resorts in the Twentieth Century*. Manchester, Manchester University Press.

Watkins, M. and Bond, C. (2007). Ways of experiencing leisure. *Leisure Sciences*, 29 (3), 287–307.

White, M., Smith, A., Humphryes, K., Pahl, S., Snelling, D. and Depledge, M. (2010). Blue space: The importance of water for preference, affect, and restorativeness ratings of natural and built scenes. *Journal of Environmental Psychology*, 30 (4), 482–493.

Williams, A. (2007). Introduction: The continuing maturation of the therapeutic landscape concept. In, A. Williams (ed), *Therapeutic Landscapes*. Farnham, Ashgate, 1–10.

Wylie, J. (2005). A single day's walking: Narrating self and landscape on the South West Coast Path. *Transactions of the Institute of British Geographers*, 30 (2), 234–247.

# 4 Sailing, health and wellbeing

## A thalassographic perspective

*Mike Brown*

## Introduction

There is a growing body of literature that has drawn on direct embodied engagement with the sea to explicate the role that the sea plays in shaping identity (Anderson and Peters, 2014; Brown and Humberstone, 2015; Peters and Brown, 2017). These works have contributed to a growing interest in how seascapes influence understandings of self, our relationships with the natural world and with others. Seascapes refer to our individual and collective consciousness of the sea, incorporating the more-than-human world, and our lived experiences of being in and on the sea (Brown, 2015). In this chapter I draw on two auto-ethnographic accounts of sailing as the basis for a consideration on how engagement with the sea plays a role in personal identity construction and can form the basis of heightened environmental awareness – both of which have implications for understandings of health and wellbeing. Auto-ethnographic accounts facilitate consideration of the fluidity and dynamism of the sea through contextualising personal encounters that link to broader social issues.

Auto-ethnography provides a methodological approach that allows us to better understand and analyse the 'connections between personal embodied nature-based experiences, culture and nature' (Humberstone, 2011: 495). As an established methodological approach, auto-ethnography incorporates reflexive self-observation along with the positioning of the researcher 'as an active and engaged participant in the social world or activity being studied' (Anderson and Austin, 2011: 2). The use of auto-ethnographic accounts in a range of disciplinary fields is well established. For example in my areas of research, outdoor learning and leisure studies, auto-ethnography is recognised as a legitimate and valuable endeavour (Allen-Collinson and Hockey, 2007; Anderson and Austin, 2011; Humberstone, 2011; Sparkes, 1999, 2009; Sparkes and Smith, 2012). Anderson and Austin (2012) have drawn particular attention to the value of auto-ethnography for exploring issues such as identity construction through the focus on the 'body and lived experience' (2012: 7). Humberstone (2015a) has also highlighted that the 'corporeal, material body is now recognised as of particular significance in many areas of social analysis, and is currently emerging in geographies of the sea' (29).

My knowledge of who I am, as a Pākehā (Maori term for a person of European descent) male, immersed in maritime matters, is inextricably linked to, and shaped by, how I perceive the world (Sparkes, 2009). As I will illustrate sailing is not only about identity formation, how I see myself, but also intimately connected to how I view my relationship with the natural world. For as Nicol has so eloquently stated, direct experiences of the sea enable '... humans to better understand their interconnection with nature' (2015b: 146).

Issues of identity, associated attitudes and the potential for developing a proactive environmental stance lie at the core of these brief auto-ethnographic accounts of sailing. The exploration of my engagements with the sea highlights the importance of direct experience in shaping a particular identity and how this impacts on my choice of leisure experiences (e.g. family holiday). As will be discussed in greater detail sailing has the potential to build an intimate relationship with the sea which carries with it, as in any genuine relationship, a duty to respond and to care. The nurturing of an ethic of care for the more-than-human world is of vital significance in a world undergoing climatic change. Failure to address current environmentally harmful practices places the health and wellbeing of humanity at increasing risk. A heightened awareness through direct engagement is one path to developing an ethic of care and development of action competence to address the growing environmental crisis (Eames, 2019). Thus, this chapter contributes to the growing body of literature that places experiences, of and with the sea, at the forefront of the contemporary field of thalassographic (sea writing).

## Narrative 1, December 2014, Atlantic Ocean

*Oh shit. Not again.*

*Leaning over the side I retch.*

*My stomach convulses.*

*Once, twice, three times.*

*My throat burns, ripped raw by acids and undigested food.*

*There's nothing left to expel.*

*It's dark and cold enough to need to wear my sailing jacket – but I'm suddenly hot and sweaty. I unzip the front of my jacket as far as I can, as much to cool down as to remove the smell of vomit that permeates the fleece lining. I know I didn't spew on it – but I can smell vomit. The odour lingers in my nostrils. It hovers on the edge of my consciousness. My jacket will only partially open – my harness chest strap acts as a barrier. I sit up, try to gather my thoughts and orient myself by the dull red light of the compass.*

*"Are you alright?"*

*Alright – oh yeah I'm great. There's nothing like being tired, sick, disoriented and asked to go on to the tossing foredeck at 3am to wrestle with a wet sail that has detached itself from the top of the mast. Half of it lies on the deck and the other half is trailing in the water acting like a giant brake.*

*"Sure, I'll be okay". And I will. I've been here before. I clip my safety tether onto the jackstay and crawl along the deck to join Andy and haul the sail aboard. For 15 minutes we struggle with the sail. The stitching and salt encrusted fabric rips the skin off my wet and softened knuckles. Each foot of sail we haul aboard is fought for. As the boat rises we pull in a few more inches of cloth only to lose them again as the boat drops into a trough. The folds overboard fill with hundreds of kilos of water and some, if not all of our hard work is torn from our hands. I'm totally focussed on the task at hand, everything beyond my restricted range of vision is in blackness and irrelevant. We shout to each other, work in unison as best we can, whilst being pitched up and down as if we're on some warped amusement park ride. We finally secure the sail and crawl along the deck back to the relative safety of the cockpit. I sit there shaking slightly. I'm breathing rapidly after the exertion. I feel worn out, my arms ache, but my sickness has passed. There is a sense of a job well done, of looking out for one another, and of keeping the boat safe.*

*I retreat to my bunk knowing that in two hours I will be back on deck for my watch.*

*At daybreak I can make out a small trail of red splashes on the white deck. Markers of my retreat from the bow hours earlier.*

Even now, thousands of kilometres and several years after this event I can still vividly recall the sensations of that night. Mack's (2007: 374) comment that the 'allure and seduction of shipboard [life] has often been described as having no equivalent ashore' brings a wry smile to my face. I certainly find sailing seductive – but clearly it is not always alluring. I have grown up around boats, helped build or refit several, worked as a professional sailor, taken my own boat on a Pacific Island cruise and now find myself managing New Zealand's leading boating education organisation. There have been extended periods away from sailing, but I inevitably find myself drawn back to the sea, either professionally or personally. In a recent paper (Brown, 2017) I explored the intersections between competence and self-identification as an offshore sailor using Ingold's (2000) concept of enskilment. The account above, drawn from early in a 21 day trans-Atlantic crossing, illustrates how I struggled to become attuned to my fluid environment in the first 24 hours of the voyage. As I explain in greater detail in that paper, given sufficient time, my body adjusted to the fluidity of movement which allowed me to regain a sense of comfort, to display competence and to feel 'at one' with my environment. I have chosen this narrative to try and reveal part of the 'reality' of offshore sailing: the initial confusion (manifest viscerally in my bodily reactions and my mental turmoil) and the unpleasant reality that is seldom seen in tourist brochures showing images of afternoon cocktails and picturesque sunsets. Despite this unpleasant opening interlude, the vast majority of my time on the voyage was far less tumultuous. The following brief narrative, recounting an event later in the voyage, illustrates how attunement to the environment permits a more relaxed and enjoyable demeanour.

> *As we accelerate white spray shots down the side of the boat as waves peel off the bow.*
>
> *I catch a glimpse of the red sleeves of my jacket at the periphery of my vision.*
>
> *There is beauty in moments like these – 'adverse weather' provides stimulating sailing. For the briefest of moments I imagine what it might feel like to be alone and charging through the Southern Ocean. I am in my element – it is not mine in the possessive sense – it is mine because I feel that this is where I am comfortable and at ease. I lightly caress the wheel, directing the boat down the face of the wave. A quarter turn to stay on the wave and then the correction to bring us back on course. These actions are instinctive – I know how far off the wind I can safely sail – how to increase the wind pressure to ensure that we get a good ride as we catch a wave.*
>
> *I'm in control. Legs braced, hands lightly grasping the wheel. I know I'm grinning from ear to ear.*
>
> *This is more like it.*

My sense of identity is linked to my perceptions of my competence as a sailor, my ability to perform the requisite roles that I perceive to be 'properties' of a competent sailor. It is an identity that I value – for it provides me with a sense of 'wholeness' – of wellbeing. I am in Fisette's (2015) terminology, embodying and 'performing identity' (86). Thinking of myself as a sailor is achieved through immersion and interaction in the activity that leads to a sense of what Goffman (1973) referred to as 'felt identity'. The account that I have presented above is a form of storytelling, which as McLean, Pasupathi and Pals suggest is an integral aspect 'to both the development and maintenance of the self' (2007: 273).

Each time I decide to go on an extended voyage necessarily entails a period of adjustment to what Casey (2009) terms a 'wild place'. Casey argues that in contrast to much of contemporary life where built environments 'serve to bring the body to rest' (2009: 223), being at sea calls forth active engagement of one's senses with a constantly moving world. For Casey,

> *wild places themselves take the lead.* However active and perceptive our bodies may be, they end up following this lead, tracing out the threads the wild world weaves before and around us. In the end – indeed, from the very beginning – we find ourselves respecting 'the lay of the land,' or the setting of the sea....
>
> (2009: 225)

The constant movement of the sea as a 'wild place' allows us to see the fluidity of this environment as more than a metaphor. As I described above, the sea's movement and changing wave patterns afford and restrict particular ways of being until one becomes attuned and skilled at living in this environment.

Part of my attraction to sailing is the requirement to be open to new experiences and the 'forced' humility of being prepared to hand over the 'lead' to wild places in order to establish a connection with the natural world. No amount of will power or self-talk will prevent me from being seasick (I've tried every trick in the book). I am required to 'serve my time' and adapt to forces beyond my control – something I am reluctant to do in my land-based life. It has been suggested that experiences such as these have the potential to enable people to learn about the world in new and more complex ways by engaging both the mind and body (Crouch, 2000). My identification with a desired self-identity and the associated sense of wellbeing is supported by a growing body of literature (I may be odd but I'm not unique!). For example the immersion of oneself in nature and natural outdoor environments has been shown to have health and wellbeing benefits (e.g. Foley, 2015; Foley and Kistemann, 2015; Kearns et al., 2014).

In the opening narrative I have chosen to present a particular story, a less than glamorous account of one experience of being at sea. I have selected this to illustrate the contingent nature of identity formation as a sailor and the power of the affective. A large body of research attests to the role that embodied affective practices play in creating a sense of self (e.g. Evers, 2006; Thrift, 2004). The affective has been described as the 'multifariousness of bodily systems which may facilitate embodied living beings' meaningful engagement with their environments, each other and non-human worlds' (Humberstone, 2015b: 64). Humberstone (2015b) argues for the value of taking into account affective narratives of being in nature as they 'tell of an extraordinary richness to people's engagements with places, landscapes and seascapes' (p. 63). She contends that it is powerful emotional experiences, processed through our senses that have the potential to engage us deeply with our surroundings. These embodied learnings in nature foster what she terms shared 'kinetic empathy' and a sense of wellbeing. Drawing on Thrift (2008), Humberstone (2015a) suggests that through movement in natural environments we might develop an empathy for environmental and social action. She notes that embodied experiences of the natural world can deepen the sense of interconnectedness of the human and non-human worlds. It is attempts to develop this sense of empathy or responsiveness that I now turn to in my second narrative.

## Narrative 2, January 2018, Tryphena Harbour, Great Barrier Island, New Zealand

*As I write this we, my partner, her son and daughter, plus the family cat, are anchored in Puriri Bay sheltering from the remnants of an ex-tropical cyclone as it sweeps across New Zealand. The current weather forecast succinctly states:*

*Marine Warning*
         *Issued at 10.58am Thursday 4 Jan 2018*
*GALE WARNING A03 FOR THE HAURAKI GULF AND THE AREA*
*FROM BREAM HEAD TO CAPE COLVILLE*
    *Northeast 35 knots gusting 50 knots tending northerly late this evening.*
*Tending westerly 35 knots gusting 50 knots Friday afternoon gradually*
*becoming 25 knots gusting 35 knots late Friday evening.*

    *Yesterday we made our way about 15 miles south from the landlocked*
*refuge of Port Fitzroy to Tryphena Harbour in order to ensure we found*
*a safe haven prior to the arrival of the expected gale. The reason for the*
*trip was to make sure that we were at our required haven in plenty of time*
*so that the kids could return to Auckland via the ferry (scheduled on 5th*
*January) to attend an extended family engagement. The approaching gale*
*meant that we arrived 48 hours early. Yesterday we walked on the beach in*
*sunshine, found a small store and bought ice creams which quickly melted*
*in the heat. We swam in the clear blue waters and could clearly see the*
*sandy sea floor five meters beneath us.*

    *We had spent the previous six nights enjoying the sheltered anchorages*
*of Port Fitzroy; swimming, walking inland to old dams constructed to*
*contain the volume of water required to transport giant Kauri logs down-*
*stream, exploring the heads of the bays and paddling up small streams*
*into sub-tropical rain forests. We had paddled with dolphins, seen a shark*
*gliding around the boat at anchor and had been entranced by the grace of*
*stingrays as they glided through the warm shallow waters in search of prey.*
*We'd eaten fish that we'd caught and carefully monitored our fresh water*
*usage to ensure we didn't run the tanks dry.*

    *It's rained constantly today and the wind isn't too strong yet it is expected*
*to build over the next 12 hours. We've played cards, spent some time cover-*
*ing material in a boating education course (buoys and beacons, give way*
*rules), eaten (surprise, surprise), read, sketched and listened to music. I even*
*managed to continue my habitual morning dip. The weather isn't looking*
*promising for tomorrow and I doubt that the ferry will be running. Even*
*if the wind diminishes the sea state will remain unpleasant for a few days.*
*I've still got another week of annual leave before I'm due back at work and*
*I'm content to relax – I learnt long ago that there's no point in fussing – the*
*weather is not something which we can control. The remnants of ex-tropical*
*cyclones are not unusual at this time of year. Two years ago we spent three*
*days in the Bay of Islands sitting out a similar storm. Perhaps I am a little*
*too much of a 'glass half full' person but I see value in days like these – we*
*get the opportunity to chat, to share what we're reading and learn the subtle*
*art of compromise in a confined space.*

Aspects of this short account may well resonate with readers – as parents of
teenagers on holiday or perhaps this reminds some of you of your own childhood holiday experiences with family. Family holidays on the boat consist

of short day sails and secure anchorages among the myriad of islands that make up the Hauraki Gulf or perhaps a little further afield in the Bay of Islands. These holidays have about as much in common with an ocean crossing as rally car driving has with commuting to the shops for groceries – both are in the car, but that is about where the similarity ends. I am careful about planning where we will go, when we will go and what we will do on arrival. This is, after all, a holiday and I am conscious of providing enjoyable and positive experiences as my partner and her children, Bailey and Thomas, did not grow up boating.

I will freely admit to, not so subtle, coercion regarding boating holidays – partly because I enjoy boating but also because I view them as a way for *us* to connect with each other and for the teenagers to develop a greater appreciation for the natural world through direct engagement. I am well aware of how much 'screen time' as adults we are exposed to and I sense that for Generation Z, those born during 1995–2010 (McCrindle, 2018), screen time is far more all-encompassing and pervasive. I am certainly familiar with the challenges that are involved in disengaging a 15-year-old boy from his phone. I'm not a Luddite, but as an educator I am concerned about the highly mediated, controlled and 'addictive' aspects of much that is available on modern electronic devices. As discussed elsewhere (Beames and Brown, 2014), what is often masked behind the commodification of experiences is a hidden curriculum that serves to normalise consumption and a disconnection from the natural world. I am also reflexive and critical enough to not fall for a Rousseauian Romanticism; one's 'real self' is not discovered in the natural world untainted by the evils of modern society. As a proponent of place responsive pedagogy (Wattchow and Brown, 2011), I, along with a host of other educators (e.g. Mannion et al., 2012; Smith, 2007), see the value of learning that is responsive to the unique places in which individuals and communities dwell.

Family holidays allow us to spend time together, largely without the distraction of electronic devices but perhaps, most importantly, they allows us to connect with the environment in ways that are more 'immediate' and direct than we often experience in our daily urban lives. On the boat our plans for today and the following few days depend largely on the wind direction and the ability to find a sheltered anchorage. A passage between islands requires patience when the wind drops, the ability to deal with minor frustrations if we take longer than expected, or accepting a change of plans depending on the weather. Being at sea also brings with it moments of immense joy and satisfaction – seeing a pod of dolphins with their calves or watching Orca swim by in search of prey. This summer provided an afternoon of excitement as Bailey paddled on a stand-up paddle board (SUP) around the bay accompanied by dolphins. Two were clearly surfacing near her, while the third 'dolphin' turned out to be a sizeable shark out for an afternoon cruise. In an instant there arose the dilemma as to alert her to its presence, and risk alarm and panic, or to quietly let it get on with its business. I chose the latter option and this led to a discussion about shark attacks, shark behaviour and the vilification of sharks (e.g. Jaws movies) in the public imagination. These conversations would not have occurred without these unplanned for, yet powerful 'teachable moments'. In several of the anchorages, we were able to quietly kayak or SUP up small streams at the head of the bay. With little imagination it felt like all that was needed was David Attenborough's voice-over to transport us into the Amazon. On the mudflats we would find stingrays resting on the seafloor and watch as they gracefully 'flew' away after being disturbed by our paddles. We watched as small fish swum in and out of the mangrove roots and discussed how mangroves provided a key ecosystem for nurturing marine life. These hitherto 'smelly mudflats' became a valued part of the land/seascape. It is in the simple things, the everyday events of life afloat that I hope the family become more attuned to the natural world around them. It is not hard to feel affinity with a dolphin, but less easy to relate to a shark – yet both deserve our respect and call upon us to consider how we live, both afloat and ashore. We certainly wouldn't throw a plastic bag overboard at sea, and we occasionally do small beach clean-ups to remove as much plastic from the high water mark as we can. Yet it is easy to ignore a piece of rubbish on the footpath as we rush to a meeting or head to the mall. Likewise we are scrupulous about what we put into the toilet or pour down the sink on the boat – as this impacts on our 'friends the dolphins', but we are less inclined to be so studious at home when waste is transported to a 'water treatment plant' and becomes someone else's issue. It is these small everyday lessons in life afloat that I hope provide experiences of significance as the teenagers' transition into adulthood.

After a lengthy career as an outdoor educator, practitioner and researcher, I appreciate the potential value of *appropriate* outdoor experiences to foster an appreciation of the natural world. It is here that my professional and personal interests mingle – I often ponder how can I provide appropriate and enjoyable experiences for my family so that together we might connect with each other and the natural world. This position, which is essentially an ethical one, was arrived at following reflecting on a comment made by Brookes and Dahle who posited that one of the most pressing educational questions of our time is 'How can and how should individuals, families, and communities experience nature in the modern world?' (Brookes and Dahle, 2007: viii). Implicit in this question is the issue of wellbeing. For how can we, individually and collectively, be well in a world that is groaning under the pressure of human exploitation?

As I have suggested in the previous narrative and elsewhere (Brown, 2015, 2017), lived experiences in and with the sea allow us to understand how it 'becomes part of us, just as we become part of it' (Ingold, 2000: 191). This effort to see ourselves as part of the world, by recognising the embodied visceral nature of our engagement with seascapes, has deep implications for individual and collective notions of health and wellbeing. Sailor and scholar

Peter Reason has drawn on Gregory Bateson, Thomas Berry and other environmental thinkers, and his own ecological pilgrimages (Reason, 2014, 2017) under sail on the western edges of Ireland and Scotland to expound the need for a fundamental rethink of how we engage with the world. The increasing acidification of the oceans and the extent of microplastics in the marine, and ultimately human, food chain is fundamentally changing our planet. The United Nations Millennium Ecosystem Assessment Report (UNESCO, 2004) provides detailed and sobering reading regarding the threats posed to the world's oceans – threats that have not reduced in the last decade as global carbon dioxide emissions and continued industrial scale fishing practices continue to place the oceans and its inhabitants under stress. As Reason explains, the resolution of these and other pressing environmental issues requires new ways of thinking. Drawing on Thomas Berry's *The Dream of the Earth* (1988), he argues that we need a 'new story' as the old story that we keep telling ourselves is no longer effective, nor does it make sense in times of rapid change. Reason embarked on an ecological pilgrimage aboard his sailing boat *Coral* in order to move beyond an intellectualising of the issues in order to break down the binary between the human and non-human worlds. As he stated, 'while our understanding of evolution and ecology tells us we are also part of the community of life on Earth, we rarely feel that in our bones or our heart' (Reason, 2017: 9). I believe that sailing provides an opportunity to feel the sea (Peters and Brown, 2017; Zink, 2015), to inhale it through one's breath (Moitessier, 1974) and to experience moments of grace (Reason, 2017). Reason's search for moments of grace, which he describes as times "when a crack opens in our taken-for-granted world, and for a tiny moment we experience a different world … It is a world … no longer divided into separate things, but one dancing whole" (2017: ix), captures the ineffable moments of oneness that can be experienced at sea.

As others (Koth, 2013; Macbeth, 1992) and I have noted, sailing provides an opportunity to interact and connect with the natural world. The importance of developing empathetic connections with the natural world through leisure experiences, either offshore sailing, coastal cruising, surfing, windsurfing or swimming, provides an opportunity to develop a relationship with a world beyond the confines of modern urban life. Such activities have, if structured appropriately, the potential to encourage behavioural change. It is clear that the impetus to find different ways of engaging with the world will gain greater importance as humanity faces changing climatic conditions (Nicol, 2015a; Reason, 2014). As suggested in both these narratives, sailing necessitates a sense of humility before 'wild places' (Casey, 2009) and also a sense of wonder.

As an offshore sailor I am required to acknowledge that, however active and perceptive my body may be, I have no choice other than to follow the lead of the sea (Casey, 2009). Likewise as the person in charge of a vessel during the family holiday 'the crew' and I need to fit our activities around the weather, which can provide opportunities for interacting in ways seldom made available at home. In sailing, we place ourselves in situations that call forth patience and the necessity to learn to live with, and incorporate the ever fluid and mobile rhythms of the sea. Perhaps it is this lack of predictability, this need to acquiesce in the face of something beyond us that we are more open to moments of grace. When these 'cracks' appear we can imagine that things can be different – it is this sense of possibility from which springs hope for a better world.

## Contemporary thalassography – writing with, rather than about, the sea

The rise of contemporary thalassography, or sea writing (Steinberg, 2014), from a broad range of perspectives attests to the role that engagement with the sea has on individual and collective identity, one's sense of place and the nature of one's relationship with the world. A focus of much of this writing is an exploration of how the sea and places located by the sea shape contemporary life (e.g. as a mode of transportation of goods in a global economy, an essential component of the biosphere, a source of food, a focus of geopolitical contestation and as the location for leisure). My intent in this chapter has been to show how by taking into account embodied experiences in the seascape we can see how our interactions shape individual and collective understandings of who we are and our relationship to the sea. Drawing on auto-ethnographic accounts, I have sought to move beyond the sea as a metaphor and engage in understanding how it 'becomes part of us, just as we become part of it' (Ingold, 2000: 191). On the deck of a sailing boat, sitting in a sea kayak or waiting for the next wave while surfing we (people of the sea) breathe the salt-laden air and feel the ocean pulse through our bodies. In these instances we become part of the fluid world that has hitherto been largely neglected in understanding the relationship between the human and more-than-human world and how this impacts on health and wellbeing.

Being on/in the sea is to enter into a relationship with it. The 'quality' or nature of this relationship is clearly open to debate. Too often the sea is seen as a dumping ground for the waste products of human activity or as a resource to be exploited (e.g. industrial scale fishing or deep sea mining). Intimate experiences of seascapes can positively impact on one's sense of self and in turn develop a sense of 'kinetic empathy' or 'place-responsiveness' (Wattchow and Brown, 2011) that may lead to pro-environmental action (e.g. Surfers against Sewage, Sailors for the Sea). Attention to embodied experiences of the sea can contribute to academic discourses of health and wellbeing. For as Evers (2006: 232) notes, 'we feel our body and then we experience emotion. Affects are what we feel at the bodily level. Both emotion and affects rely on one another to organize and register our experience with the world'. Attention to how people experience seascapes has much to contribute in humanity's quest to pursue one of the key issues facing us today – the health and wellbeing of our oceans and ultimately planet earth.

## Postscript

The final draft of this chapter was completed onboard my boat during a public holiday. Having felt guilty about 'dragging my heels' I stayed in the marina and immersed myself in the task (a task that I thoroughly enjoyed, as writing in a sustained manner is a rare treat these days). On dusk I assisted a neighbour into his berth and no sooner had he tied up than I received a dinner invitation. Freshly caught snapper, a glass of wine and entertaining company with his immediate family, friends and other guests. This spur of the moment invite, the sharing of the sea's bounty, the welcoming of friends and strangers is part and parcel of living on the marina. Is this the exclusive domain of boaties? Of course not. I feel at home in this unique gathering of like-minded souls where a sense of community prevails. These are the people who will lend you a tool, help you berth your boat, mind your pet if you're away, or pick up some groceries for you because they're 'going that way anyway'. Perhaps it's the shared experiences, the good and the not-so-good that binds us together. Or maybe it's just the simple fact that our 'homes' all move when a boat passes by and that constant movement, and too much exposure to the sun and wind, has loosened a few brain cells. Regardless I am fortunate to be in a place where material possessions take a distant second place to being on the sea as frequently as possible – what better advertisement for blue space and wellbeing is there?

## References

Allen-Collinson, J. and Hockey, J. (2007). 'Working out' identity: Distance runners and the management of disrupted identity. *Leisure Studies,* 26, 381–398.

Anderson, L. and Austin, M. (2012). Auto-ethnography in leisure studies. *Leisure Studies,* 31 (2), 1–16.

Anderson, J. and Peters, K. (eds) (2014). *Water Worlds: Human Geographies of the Ocean.* Farnham, Burlington, VT, Ashgate.

Beames, S. and Brown, M. (2014). Enough of Ronald and Mickey: Focusing on learning in outdoor education. *Journal of Adventure Education and Outdoor Learning,* 14 (2), 118–131.

Berry, T. (1988). *The Dream of the Earth.* San Francisco, CA, Sierra Club.

Brookes, A. and Dahle, B. (2007). Preface. In, B. Henderson and Vikander, N. (eds), *Nature First: Outdoor Life the friluftsliv Way.* Toronto, Natural Heritage Books, viii–xiv.

Brown, M. (2015). Seascapes. In, M. Brown and Humberstone, B. (eds), *Seascapes: Shaped by the Sea.* Farnham, Burlington, VT, Ashgate, 13–26.

Brown, M. (2017). The offshore sailor: Enskilment and identity. *Leisure Studies,* 36 (5), 684–695.

Brown, M. and Humberstone, B. (eds) (2015). *Seascapes: Shaped by the Sea.* Farnham, Burlington, VT, Ashgate.

Casey, E. (2009). *Getting Back into Place: Toward a Renewed Understanding of the Place-World* (2nd ed.). Bloomington, Indiana University Press.

Crouch, D. (2000). Places around us: Embodied lay geographies in leisure and tourism. *Leisure Studies,* 19, 63–76.

Eames, C. (2019). Developing action competence: Living sustainably with the sea. In, M. Brown and K. Peters (eds), *Living with the Sea*. London, Routledge, 70–83.

Evers, C. (2006). How to surf: Research and methodology. *Journal of Sport and Social Issues*, 30 (3), 229–243.

Fisette, J. (2015). The marathon journey of my body-self and performing identity. *Sociology of Sport Journal*, 32, 68–88.

Foley, R. (2015). Swimming in Ireland: Immersions in therapeutic blue space. *Health and Place*, 35, 218–225.

Foley, R. and Kistemann, T. (2015). Blue space geographies: Enabling health in place. Introduction to special issue on healthy blue space. *Health and Place*, 35, 157–165.

Goffman, E. (1973). *Stigma: Notes on the Management of Spoiled Identity*. Harmondsworth, Pelican Books.

Humberstone, B. (2011). Embodiment and social and environmental action in nature-based sport: Spiritual spaces. *Leisure Studies*, 30, 495–512.

Humberstone, B. (2015a). Embodied narratives: Being with the sea. In, M. Brown and Humberstone B. (eds), *Seascapes: Shaped by the Sea*. Farnham, Burlington, VT, Ashgate, 27–39.

Humberstone, B. (2015b). Embodiment, nature and well-being: More than the senses? In, M. Robertson, Lawrence, R. and Heath, G. (eds), *Experiencing the Outdoors: Enhancing Strategies for Wellbeing*. Rotterdam, Sense, 61–72.

Ingold, T. (2000). *The Perception of the Environment*. London, Routledge.

Kearns, R., Collins, D. and Conradson, D. (2014). A healthy island blue space: From space of detention to site of sanctuary. *Health and Place*, 30, 107–115.

Koth, B. (2013). Trans-pacific bluewater sailors – Exemplar of a mobile lifestyle community. In, T. Duncan, Cohen, S. and Thulemark, M. (eds), *Lifestyle Mobilities. Intersections of Travel, Leisure and Migration*. Farnham, Ashgate, 143–158.

Macbeth, J. (1992). Ocean cruising: A sailing subculture. *The Sociological Review*, 40, 319–343.

Mack, K. (2007). Senses of seascapes: Aesthetics and the passion for knowledge. *Organization*, 14 (3), 373–390.

Mannion, G, Fenwick, A. and Lynch, J. (2012). Place-responsive pedagogy: Learning from teachers' experiences of excursions in nature. *Environmental Education Research*. doi:10.1080/13504622.2102.749980

McCrindle, M. (2018). *Generation Z*. Available at: https://mccrindle.com.au/insights/publications/infographics/generation-z/, last accessed February 6, 2018.

McLean, K., Pasupathi, M. and Pals, J. (2007). Selves creating stories creating selves: A process model of self-development. *Personality and Social Psychology Review*, 11, 262–278.

Moitessier, B. (1974). *The Long Way*. London, Granada.

Nicol, R. (2015a). *Canoeing Around the Cairngorms: A Circumnavigation of My Home*. Lumphanan, Lumphanan Press.

Nicol, R. (2015b). In the name of the whale. In, M. Brown and Humberstone, B. (eds), *Seascapes: Shaped by the Sea*. Farnham, Burlington, VT, Ashgate, 141–153.

Peters, K. and Brown, M. (2017). Writing *with* the sea: Reflections on in/experienced encounters with ocean space. *Cultural Geographies*. doi:10.1177/1474474017702510

Reason, P. (2014). *Spindrift: A Wilderness Pilgrimage at Sea*. Bristol, Vala Press.

Reason, P. (2017). *In Search of Grace: An Ecological Pilgrimage*. Winchester, Earth Books.

Smith, G. (2007). Place-based education: Breaking through the constraining regularities of public school. *Environmental Education Research,* 13(2), 189–207.

Sparkes, A. (1999). Exploring body narratives. *Sport, Education and Society,* 4, 17–30.

Sparkes, A. (2009). Ethnography and the senses: Challenges and possibilities. *Qualitative Research in Sport and Exercise,* 1, 21–35.

Sparkes, A. and Smith, B. (2012). Embodied research methodologies and seeking the senses in sport and physical culture: A fleshing out of problems and possibilities. In, K. Young and Atkinson, M. (eds), *Qualitative Research on Sport and Physical Culture* (Volume 6). Bingley, Emerald Publishing, 167–190.

Steinberg, P. (2014). Foreword on thalassography. In, J. Anderson and Peters, K. (eds), *Water Worlds: Human Geographies of the Ocean.* Farnham, Burlington, VT, Ashgate, xiii–xvii.

Thrift, N. (2004). Movement-space: The changing domain of thinking resulting from the development of new kinds of spatial awareness. *Economy and Society,* 33 (4), 582–604.

Thrift, N. (2008). *Non-Representational Theory: Space, Politics, Affect.* London, Routledge.

United Nations Millennium Ecosystem Assessment Report. (2004). *Millennium Ecosystem Assessment Report.* Paris, UNESCO.

Wattchow, B. and Brown, M. (2011). *A Pedagogy of Place: Outdoor Education for a Changing World.* Melbourne, Monash University Publishing.

Zink, R. (2015). Sailing across the Cook Strait. In, M. Brown and Humberstone, B. (eds), *Seascapes: Shaped by the Sea.* Farnham, Burlington, VT, Ashgate, 71–82.

# 5    To the waters and the wild

## Reflections on eco-social healing in the WILD project

*Katherine Phillips and Antony Lyons*

## Introduction

We're standing in the shallows of the River Churn, a small tributary of the River Thames in the Cotswold Hills (UK). Late summer sunlight filters through the leaves of willow and sycamore trees that line this section of the river as it intersects fields of grazing cattle and arable crops. So far, the morning has involved cutting back and thinning the dense bankside vegetation, allowing more light to reach the water channel. Volunteers have been pulling up the invasive plant, Himalayan balsam, to provide more space for biodiversity and to improve riverbank stability. With us, in the flowing waters, are two members of the Farming and Wildlife Advisory Group (FWAG). One of us spots a crayfish, and we follow it as it slides downstream with the current. It is a North American signal crayfish, a non-native species, deemed by ecologists to be a problem in part due to the destabilising effects of its burrowing activity on riverbanks and due to its destructive impact on the smaller native White-clawed crayfish. It is an amazing creature to behold, and interacts with us as we observe it, waving its claws defensively, clearly watching our approach. The volunteer group's next activity is riverfly monitoring. We kick up the sediment in the stream, and gather samples of the turbulent water, which are then poured from buckets into shallow trays. This enables us to isolate the various types of tiny creature using a pipette, and identify the species using a chart. An electronic microscope plugged into a laptop computer allows us to see, in intricate detail, the bodies and movements of mayfly and caddisfly larvae, the latter encased in an exquisite case of minute, multicoloured sand grains. Enlarged on the screen, the movements of these creatures are mesmerising, as though performing an intricate dance. The day is rich with such multispecies encounters.

Led by FWAG, the WILD (Water and Integrated Local Delivery) project operates on the whole catchment, or watershed, of the River Churn. Working with volunteer groups, farmers, residents and local organisations, WILD takes an integrative approach to enhancing water quality and ecology, attending to social and policy rifts and disconnects, as well as individual healing and wellbeing. Our initial contact with WILD was made via the production process of a short documentary film with an environmental communication purpose.[1] In this exploration, we revisit the semi-structured interviews conducted for the film-making – with coordinators, volunteers and others – in order to further explore the dynamics of the project that seemed to us to interweave environmental and social healing, through what we term 'eco-social healing'. In doing so, we contemplate the significance of water (the 'W' in WILD), as a medium of transformation at multiple levels. Drawing also on our own experiences with WILD we highlight some significant elements of eco-social healing, including the multisensory and multispecies encounters. In the second part of the chapter, we reflect on the model of 'integrated local delivery' and its relationship to land-use management and people's diverse engagement as 'hydrocitizens'. We consider how the approach taken by WILD can provide a holistic ecological model, enabling an integration of healing and recovery on individual, social, community and bioregional levels.

We use the WILD case study to forefront ideas which have been emergent for us in other spheres of socioecological research, especially within an academic project entitled *Towards Hydrocitizenship* (2014–2017)[2] which creatively examined multiple facets of a water-oriented 'ecological citizenship' (Dobson, 2007). While making reference to the links between environment and health, we seek to move beyond an anthropocentric view of the health benefits provided by the environment for humans, following streams of thought into the fertile realms of interactions between human and more-than-human in the broadest sense (other animals, plants, biological systems, water, etc.).[3] We draw inspiration here in particular from the Anthropocene-critiquing provocations in the work of Donna Haraway (2016). We explore some therapeutic ramifications of multispecies and multisensory encounters. We find parallels with the ethnographic work carried out by Strang (chapter two in this volume), referring to the water-related cultural symbolism and attachments of aboriginal people in Cape York, Australia, and with David Abram's (1996), explorations into ways of relating with the natural world that are more expansive, inclusive, integrated and respectful than those which have become dominant in the contemporary world. In our estimation, holistic projects like WILD have some resonances with non-Western cultural scenarios that exhibit and value strong bonds between humans and their environments.

Our interest in water includes an awareness of the dual physical and symbolic nature of river/water circulation networks (Illich, 1986) and their intrinsic qualities of mesh-like connectivity and unboundedness. With their organ systems, living animal bodies echo such circulation flows and functionalities. In an awareness of these resonances, we consider modes of recovery and recuperation that can span biophysical systems as well as social groups and the individual. Just as Guattari (1989) emphasised the concurrence of three registers of 'ecology', namely social, mental and environmental, our exploration of 'eco-social healing' in WILD speculates on the role that an integrative framework of activity can have on these different yet interconnected registers.

## Part I: going WILD by the river

It is late autumn. There is a steady downpour of cold rain and everything is sodden. We are out on another 'river restoration day'. Three groups of workers, from the WILD project, the local authority groundwork team and the Gloucestershire Wildlife Trust, are working together to construct bat boxes and lay the foundation for a new footbridge over the river. In spite of the challenging conditions, everyone seems cheerful and involved. A group of butterfly enthusiasts gather everyone together and give a talk (as the rain carries on in an unrelenting drizzle) about how to identify the eggs of butterflies and protect their habitats when removing vegetation. Though the volunteers are keen to return to their active work, they take time out to listen and add to their knowledge of the fellow inhabitants of the local environment, this time of the winged variety. In the assemblage of the river, the future butterflies and moths, the prospective bats and the sodden volunteers, we are embedded in a multispecies "meshwork of entangled lines of life, growth and movement" (Ingold, 2011: 63).

### *Expanded ideas of health in nature*

Research in health geographies and related disciplines has tended to focus on the individual and to a lesser extent on collective therapeutic effect of 'exercise' interactions in/with nature such as walking or swimming (e.g. Barton and Pretty, 2010; Foley, 2017). In the field of environmental psychology, Kaplan (1995) proposed the idea of 'attention restoration' based on studies showing the impacts that activities in nature had on people's ability to perform set tasks, while much research now shows a positive correlation between time spent in nature and a person's overall health and wellbeing (e.g. Abraham et al., 2010). Such work has resulted in recent policy shifts towards valuing nature for human health, both in health policy (Shanahan et al., 2015) and also those relating to land use and management, with advocates of 'ecosystem services' assessing and quantifying the value to human health of access to natural spaces. Yet such policy or political aspirations are critiqued for their narrow focus on metrics that deal with the natural world and measure human interactions in a fairly mechanical and reductionist way.

In contrast, if we follow lines of thought set out by Haraway (2016) and others (e.g. Morton, 2010), what is needed is to recognise our embeddedness, or immersion within the natural world, rather than to view the environment as something external and separate from us. This accords with Morton's idea of 'the ecological thought' (2010), and the post-human feminist phenomenology of Neimanis (2017). What they are impressing on us is that there is no real separation between the ecological, the social and the individual.

We argue that certain modes of group participation in nature-based work – especially involving water as central – can provide situations where one feels 'flow' and therefore healing and recuperation. The intrinsic qualities of water (and water environments) disrupt by introducing flow and unpredictability. Water connects with, and through, all life, while able to dissolve barriers and rigidity, breaking out of every constraint. Water operates on humans in psychological and physiological ways. Following Neimanis (2017), we emphasise the correspondence, and continuum, between the nature of geographical water connectivities and the make-up of the human organism. In the context of the WILD 'river restoration days', we ask whether experiencing altered modes of connection, including with fellow humans through shared working in/around water, may be seen as an aspect of 'the hydrocommons of wet relations' (Neimanis, 2017: 6).

### The multisensory experience

Having reflected on the particular qualities of water, its mutability, its permeation through all life and its unboundedness, we feel it makes a particularly effective conduit for the healing of disconnections.[4] The fascinating qualities of water and our deep reliance on it for life are perhaps also encapsulated in the common propensity to enjoy its sensory aspects. We are drawn to the sonic qualities of water, for example the rhythmic swoosh of waves on a shore, the burbling of a brook or the energetic crashing of a waterfall. In blue space research, these sonic properties have been identified among other qualities as part of a salutogenic health effect (Völker and Kistemann, 2011). To assess the attraction of the immersive, bodily experience in water, one needs only look to coastal settings in hot weather, to see how they become inundated by human hordes paddling in the shallows, surfing waves or immersing fully for a swim. The visual qualities of water also provide a focal point in a landscape, and are a source of emotional sustenance as well as being strongly linked to emotions about place (e.g. as noted by interviewees in Strang, 2004).

In our discussions with the WILD participants, the sensory aspects of water were often mentioned. For instance, the soundscape of water was placed in opposition to a soundscape of traffic by Archie, the volunteer coordinator who ran the river restoration day:

> Last week where we did the riverfly monitoring, standing in the river – very therapeutic. It is gorgeous. Because sometimes we have a cup of tea near the river and all you can hear – you can't hear traffic – all you can hear is the water. Absolutely beautiful.
>
> (Archie, volunteer coordinator)

For Archie, and for many involved, part of the attraction of the activity is an altered sensory experience. Similarly, Helen Richards, community engagement officer with WILD, described how being near water is relaxing, partly due to the sound, and partly due to the visual sense of a continually changing waterscape of light, colour and reflections. A river might vary from rushing and full after heavy rain to a quiet, almost still flow at drier times. The change means that the same area or body of water is always interesting, and repeated interactions with it allow for a further sense of connection. Volunteer coordinator Joanne Leigh also spoke about the attractiveness of water, noting that in a landscape, or on a walk, water adds diversity and interest, whether its puddles that children (and some adults) like to splash in, a stream to jump across or looking in water to see what creatures live there. She described the delight of those involved in the riverfly monitoring at discovering a whole miniature world teaming with life, and how this captures the imagination. Both Helen and Joanne talked about the sense of fun and interest, the playful aspects of interactions, getting muddy, perhaps making a swing over a river or building a dam in a stream, and experiencing a diversity of sensations, sights, sounds and interactions with phenomena like the flow of water, the texture of mud and the discovery of fellow creatures living in those environments.

In *Towards Hydrocitizenship*, we contemplated the way in which water acts as a conduit or connective force for people's sense of themselves (as well as a sense of connectedness with other humans and with the non-human material world). The diversity of embodied experience in the interactions with the river environment can be seen to mirror the diversity of human life and experience, as has been discussed in blue space geographies (e.g. Foley and Kistemann, 2015). The changing environment speaks to a sense of flow through all things, and connectivity between all things. The plants, creatures, river banks, fields, livestock and people are all part of this interconnected flow, and without necessarily making this explicit in a conscious way, the WILD project brings awareness to this through the embodied interactions.

### Purposeful activity and healing disconnections

Besides the multisensory aspects of interacting with water, the volunteer activity on the river involves another important element: that of purposeful physical work requiring absorbed attention. Whether it is cutting back branches of trees, pulling up Himalayan balsam, building structures such as bridges, willow spiling (bank stabilisation) or gravel ramps for cattle watering, the WILD volunteers are generally involved in some form of active work. In the most simplistic sense, it could be argued that these forms of exercise have beneficial effects for health, causing elevated heart rates and deeper breathing. In the UK, diseases caused by sedentary lifestyles are on the rise, and accessible physical activities are of clear benefit in this respect (NICE, 2018).

However, there is more to the studied volunteer activity than simply movement. The tasks are purposeful, focused on improving the ecological health of the river, and integrated with the management of the surrounding land. They also contain elements of both individual and cooperative working. One person may be tackling a section of vegetation as a solo piece of work, but techniques and knowledge are shared and tools and equipment are handed around. And of course, the daily rhythm is punctuated by collective tea breaks. Although simply exercising might result in similar beneficial effects, the purposefulness and meaningfulness of these activities, the ability to instantly see the results of the work, the sense of self-worth and the social nature of the work all contribute to mental wellbeing. This resonates with Antonovsky's influential idea of salutogenesis, in particular relating to meaningful work. There is a mode of reconnecting body and mind, or body and spirit, through doing work that will have broader community and environmental benefit. Similar findings have been reported in other projects involving volunteers or citizen scientists in environmental care (e.g. Koss, 2010), and in relational approaches to wellbeing and sustainability (Helne and Hirvilammi, 2015).

The beneficial effects of this kind of work on both physical and mental health have been recognised by healthcare practitioners, and seized upon by those looking of ways of decreasing the burden on public healthcare. 'Social prescribing' is an emerging approach that involves prescribing or encouraging those affected by physical and mental health issues related to sedentary and isolated lifestyles to try social and community activities, such as volunteering on outdoor projects (King's Fund, 2017). On another river action-day with a WILD group, we interviewed someone who has come on a journey via social prescribing. Some six or seven years before this interview, Shaun had been struggling with physical and social problems in his life until he encountered the local Green Gym[5]:

> I was somebody that never went out, never went and saw anybody. Just stayed at home, watching TV. Go down, claim my giro, go back home, and that would be it. I'd moved away from the people I was involved with … so I had no social circle. So that was literally, that was it, that was my life. So I found out about the Green Gym … [And because] it was all age ranges I thought it can't be that hard … Archie's always welcoming, and he's really easy, so he made it comfortable, and comfortable's what I needed at the time. And then the more I did it, the further I got, the more I wanted to do it. Archie and doing the Green Gym helped me, you know, develop the people skills but also the physical ability, so you know, it did turn my life around.
>
> (Shaun, Cirencester County Council grounds team manager)

From there, Shaun joined in further activities, and when we met him he was managing a small groundworks team from the local authority and in the midst of helping to build the foundations for a footbridge. Shaun was clearly delighted at how his life had been transformed and, locally, he has become something of a poster boy for social prescribing.

Although not every story of eco-volunteering is quite as transformative as Shaun's, our interviews and observations reveal a situation where individuals are benefiting personally from the activities in the groups, through developing their sense of achievement and self-worth or value. In certain cases, the volunteering has provided distraction from traumatic memories or from rehabilitation difficulties encountered by former military personnel, or those who have been using drugs as a coping mechanism. It provides meaningful social activity for individuals who may otherwise be in situations of social isolation and unstructuredness. There is a fluidity and rhythm to the group activity, a pattern of working interspersed with cups of tea provided by Archie from the smoking, steaming 'Kelly Kettle'[6] fuelled with twigs gathered from the surroundings, itself an elemental and symbolic process that enhances connectivity to the environmental setting. The group social environment that is fostered is one of mindful attention and 'communitas'. For Turner (1969: 95) communitas exhibits a sense of equality where 'secular distinctions of rank and status disappear or are homogenised' and 'communitas often appears unexpectedly. It has to do with a sense felt by a group of people when their life together takes on full meaning' (Turner, 1977: 1).

One of the essential characteristics to the building of 'communitas' within the work groups hinges on sensitive coordination by the facilitators. There is a strong culture of acceptance of everyone as they are, as they come, with support, encouragement and enthusiasm for all degrees of contribution. Volunteer coordinator Joanne talks about how, to an extent, she designs activities around the kinds of things that participants want or need to be doing, working to balance this fluidity and sensitivity with the broader objectives of the project, i.e. enhancing the ecology and quality of the river environment.

## Part II: water and integrated local delivery

A key aspect of the WILD project is the fact that it is organised around the central element of water. Water is what provides the platform and geographical milieu for 'integrated local delivery', or in other words, water is what inescapably connects everyone and everything in a local (and global) environment, and as such it is an appropriate medium through which to develop and uncover connections, and to forge and foster relationships. As Joanne Leigh, one of the volunteer coordinators, says,

> We can appeal to landowners and farmers because we've got that direct link to them. But what about the rest of the people? How do you link those to the land that they live in as well … but [a] tiny little speck of land could make all the difference to the quality of the water? So yes, this has veered to parishes and managing water through the parishes by fixing ditches, by fixing culverts, and that sort of thing, and I guess by doing that you're engaging the people of the community who are then looking further out to the land around them, getting a better relationship

with the people who own the land around them, going back to the way
we used to be 70–100 years ago, where everybody worked on the land,
rather than, not – in the city, with no relationship to the land at all.

(Joanne Leigh, volunteer coordinator)

Joanne refers here to impact of micro-care efforts on lessening flooding by
careful clearing of ditches and so forth. There is a health side effect of this
in terms of environmental risk and the environmental management efforts
can therefore be seen as a form of 'preventative health' intervention. The essence of WILD isn't simply about the river being healed or individuals being
healed; rather, it's about the holistic quality of a locality and the reintegration
of the elements that comprise such a locality, including the humans, animals,
insects, plants, water, soil and so on. As Joanne notes, the approach of the
WILD project has much to do with an integrative thinking and approach.

The 'integrated local delivery' core of WILD is premised on the idea of involving not only regulatory bodies such as the Environment Agency (alongside
other statutory authorities) and land managers, but also – crucially – local
people in multiple ways of attending to their local environment, with a primary
focus on water. A special aspect of the WILD format is the role of a dedicated
local adviser who works with stakeholders to initiate a comprehensive and accessible mapping or meshing of systems of environmental management and
improvement. In developing such links, to reconnect people with their watery
environments, this intermediary or facilitator is crucial:

> You do need somebody that can inspire people to feel that connectivity
> with where they live and their environment, and then they can become
> part of something really meaningful and feel valued … And that's the
> bit I love about these projects. They actually break down barriers and
> they create this lovely opportunity for people to become resilient by
> working together and valuing each other's role in society.
>
> (Jenny Phelps, WILD coordinator)

Part of the culture of acceptance and appreciation of others has to do with
an attitude towards knowledge. Integrating activity also means integrating
knowledge or understanding between, for example, ecologists, civil engineers, flood managers, water quality scientists, farmers and land managers
and a wide variety of other people who live locally. In interviews, it was
suggested that various experts might find it difficult to look at things from a
more holistic perspective:

> I think acknowledging that you don't know everything is part of it. It's
> almost like acknowledging if you're ill and you need to get help … And
> I suppose that's what the whole framework's about. It's about sort of
> realizing that we've all got that contribution to make.
>
> (Jenny Phelps, FWAG)

So yes, we can pull in specialists in the Environment Agency ... but there are people in these communities that have got an awful lot of experience and technical knowledge as well and it's respecting that and helping them understand maybe how things have changed, and why we have to do things differently. But not treating them as a layman that knows nothing. And no one does in the project, that's the lovely thing. We value everyone's opinion. And their feelings. And we let everyone have their say. And then we try to piece things together.

(Helen Richards, community outreach officer for WILD)

In our ongoing contemplation of the idea of 'hydrocitizenship', we see the WILD project as embodying the kind of meshing role that allows for people as well as the non-human to arrive with different contributions and discover a sense of belonging in relation to water. There are resonances here with a growing sense of the eco-social in social work (Norton, 2012), wherein the wellbeing of an individual is beginning to be seen in relation to their contact with the wider community not only of people but of the environment.

The impetus for all the work that FWAG does on the WILD project derives from the Water Framework Directive (WFD), which is EU legislation designed to improve water quality, habitats and biodiversity (though there are also directives specifically relating to habitats and biodiversity). As well as setting standards, the directive seeks to involve a wide array of affected and interested parties in the delivery of better quality water. Indeed, one of the two core aims of the WFD is 'getting the citizens involved' (European Commission, 2016), based on a sense that citizen involvement is crucial in order for the environment to be cared for effectively. The WFD captures both the complexity and simplicity of water management or water governance. In its simplest form, it is about achieving better chemical and ecological water quality, and getting citizens involved in this imperative. Yet addressing this aim is a highly complex process. The WILD project, with water at its core, often needs to grapple with complexity and engage with people, relationships and places. As Joanne Leigh puts it,

The fact that you've got polluted rivers, rivers that are failing for fish populations, or silting, or all sorts of things that have been passed down to us from the Environment Agency and from the Water Framework Directive, and that was the bit we planted our project on, and then worked back from the river, so we're trying to fix everything that ends up in the river so that becomes, more pure I guess, the water becomes better quality water ... that's the core of the project, but all the work we do, is away from the water. It's based around water bodies ... but it's about the land management.

(Joanne Leigh, FWAG)

People think that the [WILD] framework is really complicated. It's not complicated. It's really really simple ... Because we're all part of it. We're all part of a society. We all have a role to play, and that society relies on the environment to protect us, like any other species, you know we require food, water, shelter and the ability to exhibit natural behaviour... I think the challenge is for people to see that you're part of a system, a natural system.

(Jenny Phelps, FWAG)

There are many features about the WILD approach that are inspiring. The work with parish councils and volunteer groups is a step towards reconnecting people with the natural and social worlds of which they are part. In diverse ways, this is a project of eco-social healing.

## Conclusions

In considering how the activities of WILD can be seen as eco-social healing, we have touched on the importance of the meshwork of multispecies interactions, and the physical and psychological aspects of water that engender wellbeing. We revisit Guattari's three ecologies and see positive aspects of belonging, engagement and wider responsibility (or 'response-ability' as per Haraway) in the WILD project, enabled through the particular qualities of water.

We have taken a fluid approach to exploring the WILD project, echoing the unbounded quality of water. From our observations and interviews, we derived a sense of a project replete with woven interconnections, of welcoming ideas and contributions from all participants and piecing these together. In our transdisciplinary approach to assembling our reflections, we have adopted ideas and inspirations from writings in environmental psychology, eco-feminism, eco-social work, sustainability studies and health geographies, among others. The ubiquitous and mutable quality of water, and its permeation through all life and matter on earth seem to demand this kind of braided thinking, offering a challenge to the kind of reductionist and instrumental approaches that have arguably been causative of our gradual disconnection from natural systems. At the confluence of different disciplines is a space that has allowed us to consider the links between individual health, social health and environmental health, as always emergent interactions of mutuality and reciprocity.

The WILD project strikes us as a good demonstration of hydrocitizenship. Through multisensory and embodied interactions, connection with self and place is developed. We see a relationship between environmental volunteering and shared care between human and more-than-human. We observed the significance of being involved in work that is meaningful and purposeful work in relation to water, and the impacts of this on sense of self-worth and belonging. The framework or paradigm upon which WILD is based is one of recognition that everyone brings a contribution in relation to water in the environment, and thus a sense of connectivity centred on water permeates the activities of WILD and enables (re)connection on multiple registers or levels. As such, we feel that WILD is a component of a paradigm shift. Through fluidity and sensitivity to tensions and opportunities – for people and places, WILD shows a path towards recovery on a damaged planet.

## Notes

1 The film and further information about WILD can be found at www.fwagsw. org.uk/wild-project.
2 More information about *Towards Hydrocitizenship* can be found at www. hydrocitizenship.com.
3 Following prior explorations with water as part of the more-than-human participatory research project: http://www.morethanhumanresearch.com/conversations-with-the-elements.html.
4 This is not to overly romanticise the connective possibilities of water. After all, water can be a cause of divides and disconnections as much as the converse, as illustrated by bitter disputes over rights to water bodies for irrigation or other uses, particularly when political boundaries are involved. However, for the purposes of this chapter we are interested in the quality of connection that water manifests and the possibilities contained within that.
5 An initiative of social prescribing runs from Cirencester hospital involving volunteer groups to clear overgrown vegetation from an old orchard, plant new trees and put in pathways and benches for hospital inpatients to enjoy.
6 A Kelly Kettle is a simple device used to boil water by burning small twigs in a central cavity around which the metal body of the cylindrical kettle allows maximum heat to transfer to the water.

## References

Abraham, A., Sommerhalder, K. and Abel, T. (2010). Landscape and well-being: a scoping study on the health-promoting impact of outdoor environments. *International Journal of Public Health,* 55 (1), 59–69.

Abram, D. (1996). *The Spell of the Sensuous: Perception and Language in a More-Than-Human World.* New York, Vintage Books.

Antonovsky, A. (1979). *Health, Stress and Coping.* San Francisco, CA: Jossey-Bass.

Barton, J. and Pretty, J. (2010). What is the best dose of nature and green exercise for improving mental health? A multi-study analysis. *Environmental Science & Technology,* 44 (10), 3947–3955.

Dobson, A. (2007). Environmental citizenship: Towards sustainable development. *Sustainable Development,* 15, 276–285.

European Commission. (2016). *Introduction to the New EU Water Framework Directive.* [online]. Available at: http://ec.europa.eu/environment/water/water-framework/info/intro_en.htm, last accessed July 22, 2018.

Foley, R. (2017). Swimming as an accretive practice in healthy blue space. *Emotion, Space and Society,* 22, 43–51.

Foley, R. and Kistemann, T. (2015). Blue space geographies: Enabling health in place. *Health & Place,* 35, 157–165.

Guattari, F. (1989). *The Three Ecologies.* Translated by I. Pindar and Sutton, P. London, New Brunswick, Athlone Press.

Haraway, D.J. (2016). *Staying with the Trouble: Making Kin in the Chthulucene.* Durham, NC, Duke University Press.

Helne, T. and Hirvilammi, T. (2015). Wellbeing and sustainability: A relational approach. *Sustainable Development,* 23, 167–175.

Illich, I. (1986). *$H_2O$ and the Waters of Forgetfulness.* London, Marion Boyars.

Ingold, T. (2011). *Being Alive.* New York, Routledge.

Kaplan, S. (1995). The restorative benefits of nature: Toward an integrative framework. *Journal of Environmental Psychology,* 15, 169–182.

The King's Fund (2017) *What Is Social Prescribing?* [online] Available at: https://www.kingsfund.org.uk/publications/social-prescribing, last accessed July 20, 2018.

Koss, R.S. (2010). Volunteer health and emotional wellbeing in marine protected areas. *Ocean & Coastal Management,* 53 (8), 447–453.

Morton, T. (2010). *The Ecological Thought.* Cambridge, MA, London, Harvard University Press.

National Institute for Health and Care Excellence (NICE). (2018). *Physical Activity and the Environment Guidelines* [N90]. [online]. Available at: https://www.nice.org.uk/guidance/ng90, last accessed July 22, 2018.

Neimanis, A. (2017). *Bodies of Water: Posthuman Feminist Phenomenology.* London, Bloomsbury Publishing.

Norton, C.L. (2012). Social work and the environment: An ecosocial approach. *International Journal of Social Welfare,* 21 (3), 299–308.

Shanahan, D.F., Lin, B.B., Bush, R., Gaston, K.J., Dean, J.H., Barber, E. and Fuller, R.A. (2015). Towards improved public health outcomes from urban nature. *American Journal of Public Health,* 105 (3), 470–477.

Strang, V. (2004). *The Meaning of Water.* Oxford, Berg.

Turner, V.W. (1969). *The Ritual Process: Structure and Anti-Structure.* London, Routledge & Kegan Paul.

Turner, V.W. (1977). *The Ritual Process.* Ithaca, NY, Cornell University Press.

Völker, S. and Kistemann, T. (2011). The impact of blue space on human health and well-being–Salutogenetic health effects of inland surface waters: A review. *International Journal of Hygiene and Environmental Health,* 214 (6), 449–460.

# Part II

# Experiencing health in blue space

# 6 From water as curative agent to enabling waterscapes

## Diverse experiences of the 'therapeutic blue'

*Karolina Doughty*

## Introduction

Recent years have seen an increase in scholarly attention applied to the experiential relationship between humans and water. Significant insights have been gained into the human-water relationship more broadly, for instance in regard to the rich and evolving meanings of seascapes (Brown and Humberstone, 2015), as well as the growing literature on the health-enabling potential of being in or near water (Foley, 2010, 2011, 2014; Foley and Kistemann, 2015). In relation to questions about water and health, the literature within and beyond health geography exploring 'therapeutic blue space' has emerged strongly, contributing to the already large body of work which has applied the concept of therapeutic landscape (Gesler, 1992) to a wide range of contexts, to investigate how environmental, societal and individual factors interact in the creation of health-enabling places (for a scoping review, see Bell et al., 2018). In Gesler's (1992) original conceptualisation, a therapeutic landscape is a place (a) where a material setting has been created to support the pursuit of health and wellbeing, (b) which is culturally associated with health and (c) where social practices related to 'healing' take place. Through these three elements the 'healing process' is situated geographically in places. As such, the therapeutic landscape concept has been applied to a wide range of environments from the perspective of exploring the attribution of health-related meaning to places and landscapes by individuals, groups and more broadly societies.

As the research on therapeutic landscapes has developed, it has contributed rich insights into varying and blurring 'palettes of place' (Bell et al., 2018), where green spaces have long been central within research on the potentially beneficial interactions between place and health, but where other 'palettes' are now emerging, including a growing literature on 'blue' spaces. A number of recent accounts have highlighted the lived meanings and practices through which the particular materialities of a broad range of settings featuring water become associated with healing or health promotion (Bell et al., 2015; Foley, 2015; Foley and Kistemann, 2015;

Kearns et al., 2014; Lengen, 2015; Thomas, 2015). These studies have dealt with a range of settings where engagements with water are perceived to promote healthy living, such as islands, cities, rivers, coasts, beaches and lakes, and a variety of practices that mobilise and harness these effects, such as swimming, promenading, retirement and walking (Bell et al., 2018: 124). This growing literature has begun to build awareness of how blue landscapes serve as important sites of meaning and health-related practices for different people, at different times, or under different circumstances. The notable literature on 'blue spaces' has more recently been complemented by accounts of hybrid environments such as the green-blue landscapes (Finlay et al., 2015; Foley, 2015) or the grey-blue of water in urban spaces (Völker and Kistemann, 2015), as well as the complementary expanding palettes of brown and grey of built environments such as allotments or community gardens (Pitt, 2014), and the white of winter landscapes (Finlay, 2018). The growing focus on affective, embodied and multisensory outdoor experiences both within and beyond the therapeutic landscape literature (e.g. Spinney, 2006; Straughan, 2012) has also highlighted the risk of neglecting the wider textures of therapeutic landscapes when framing such spaces primarily through colour (Brown, 2016).

This chapter builds on this existing research by considering the contribution that health geographies can make to the study of water and wellbeing by emphasising alternative measures of value for watery spaces, particularly through experiential approaches to the concept of therapeutic landscape, that highlight the diversity of such spaces. In this effort, the chapter engages with the shifting range of ideas and practices that connect water with health and wellbeing, tracing the development of ideas which have positioned water as 'curative agent' to more recent understandings of water as part of a wider health-promoting environment, mobilised through embodied and sensory engagements on an individual level. The chapter highlights that an understanding of blue space as a potentially 'enabling' environment for human wellbeing needs to be situated both within the embodied and sensory particularities of therapeutic place-engagements, and at the same time respond to broader discourses that frame personal health and wellbeing. It is within the growing contemporary popularity of wellness and lifestyle, after all, that blue spaces have gained so much interest for health-seeking over the past decades (Kruizinga, 2016: 391).

## The shifting conceptions of the therapeutic blue

Throughout history water and its use have reflected various experiences, values and interpretations about wellbeing, health and illness (Verouden and Meijman, 2010). The history of meanings and representations associated with water and its potential salutogenic benefits can be seen to have shaped, and been shaped by, a host of different interactions with water – be it practices of bodily immersion, the drinking of sea or spring water, or simply being in proximity to water. This part of the chapter provides a brief overview of the medical and popular thought and practice which has associated water with healing effects, from the 18th century to the present (Charlier and Chaineux, 2009; Foley, 2010; Kruizinga, 2016). In early conceptualisations, the therapeutic potency of natural water was thought to lay specifically in its chemical and physical properties, thus framing water as a 'curative agent' able to infer healing of existing ailments, shifting slowly along with a broader medicalisation of everyday hygienic practices to be seen more as a medium for health-promoting practices. Through this broad development of ideas and practices, we can see shifts that are indicative of changes in medical knowledge and discourse, as well as changes in the positionings of the cure-seeker themselves, between the poles of 'passive patient' and responsibilised neoliberal subject.

### *Water as curative agent*

As Linton (2010: 3) put it, 'water is what we make of it'. Indeed, ideas and practices associated with water are context-dependent; water has been assigned different meanings, characteristics and qualities through the ways that we have engaged with it through different historical periods and in different cultural, social and political contexts (Lykke Syse and Oestigaard, 2010). We can follow the significant history of ideas and practices that valorise water as a curative agent from the thalassic therapies of ancient Egypt, France, Italy and Greece, to the emergence of seaside resorts across Europe in the 18th and 19th centuries, to the culture of the spa as a space for rejuvenation and restoration that has persisted into modern times. The idea that immersion in, or the drinking of, natural waters, cold or heated, was connected to improved physical and mental health re-emerged strongly in the 18th century in the West (Verouden and Meijman, 2010). Taking the waters and air of the seaside, or receiving the 'water cures' offered by inland spas, became increasingly established as curative practices in Europe through the late 18th and 19th centuries up until the early 20th century. As holistic therapeutic landscapes, bathing resorts and spas offered opportunities for healing, recreation and socialising to varying but mutually complementary degrees. It was believed that the medicinal value of the natural waters required continuous exposure, and thus stays of several weeks were legitimated, which contributed to spas also becoming vital spaces of leisure, which functioned as 'arenas of play and social hospitality and interaction' (O'Dell, 2010: 39), although reportedly spas in France were designed for therapy not leisure, and taking the waters was 'a serious activity and quite sober' (van Tubergen and van der Linden, 2002: 274). Spas during this period exemplify the 'traditional healing sites' theorised by Gesler (1992) as enabling physical, psychological and spiritual healing through their unique complementarity of place-based meanings (such as the symbolic meaning of water and the sense of place of the spa as a setting) and wider sociocultural context relating to health (medicalisation of society, the moralisation of hygiene, the fashionability of the spa stay, etc.).

The application of seawater in medical circumstances was elaborated by a number of physicians across Europe during the 16th and 17th centuries. As early as in the 16th century, Italian doctors recovered lost texts on medical treatment using water from the ancient world, and balneotherapy again came under consideration by the medical community, resulting in a number of publications – among them were an encyclopaedic work containing an overview of ancient and modern literature on the use of medicinal water; *De balneis omniae qua extant*, published in 1553; and *De Thermis* by Bacci, published in 1571 (van Tubergen and van der Linden, 2002). During this time, the first attempts to analyse the waters for their mineral components were made; 'it was equally important to recognise the quality of each mineral and its effect on the body, as to know which parts of the body might be influenced by taking the waters' (van Tubergen and van der Linden, 2002: 274). By the 17th century, this new bathing culture had spread across Europe, and was particularly popular with the elite. In England, John Floyer (1649–1734) published a groundbreaking work in 1697 on the history of cold bathing, which enthusiastically emphasised the curative properties of water (Verouden and Meijman, 2010). Another noteworthy medical treatise on the health-giving properties of seawater came in the mid-1700s by Dr Richard Russell (1687–1759) of Brighton, who published a widely acclaimed work on the use of seawater to cure glandular disease. Although disagreements existed about how to best use seawater to harness its curative power, when it came to the curative properties themselves there was wide-held agreement, embraced by a number of English and French physicians during the mid-18th century, that seawater was indeed beneficial for health (Gray, 2006). During the 18th and 19th centuries, ideas about the therapeutic potential of sea- and mineral-water's chemical and physical properties shifted, and ways to harness this power in treatment applications were increasingly developed; 'The specialty of balneology and hydrotherapeutics that developed from the 1880s engaged with new cultures of scientific analysis and systematic investigation within medicine using chemistry and physiology to assess and measure the healing potential of waters' (Adams, 2015: 3). In this early modern period, disease and ailments were still largely understood as arising from the relationship between the individual organism and the environment, with cures often involving the restoration of harmony between the body and nature (Verouden and Meijman, 2010). Any clean water could be used, but cold spring water especially was believed to stimulate the body's innate power to heal itself (Adams, 2015). Beneficial effects of natural waters were outlined for an astonishingly diverse range of illnesses across Europe. For example, Dr Hellman of the Ronneby health-well in Sweden reported in the 1860s the effective treatment of a broad spectrum of ailments including rheumatism, epilepsy, psychological disorders, hypochondria, hysteria, gastric disorders, spasms, gonorrhoea, bronchitis, worms and cardiovascular problems (O'Dell, 2010: 39). Prescribed treatment plans involving bathing in and ingesting natural waters often featured highly specific regimens, developed to harness the curative effects stemming from the unique properties of local waters, be it seawater or natural springs. For all the contingent pleasures and enjoyments of the spa environments of the 19th century, the water cures themselves were not always pleasant or comfortable, but rather required a measure of moral courage from patients (Verouden and Meijman, 2010). During this time, there was a perception of water as a symbol of discipline and order, and a number of high-tech water treatments were developed – such as cage showers and mechanical sprays – through which spa doctors took control of water; 'they used their power over water as a symbol to emphasise and assert their control over the outside and the inside of the body' (Verouden and Meijman, 2010: 24). O'Dell (2010) reports the example of a woman who arrived at the Ronneby health-well spa in Sweden in the 1860s with complaints of weak nerves; she was ordered by the attendant physician to ingest three-and-a-half mugs of mineral water per day, gradually increasing the dose to five or six. In addition, she was to bathe in iron-rich water heated to 35 degrees Celsius for five minutes every third day. The medical records note that after a six-week stay at the spa, the young woman was cured of her ailments. Another example of a success story from the Ronneby spa involved a man who was suffering from paralysis of the right side of his body, perhaps after a stroke; after 27 days of ingesting increasing quantities of water he was said to be able to stand on his bad leg, and by the 36th day the records state that he was able to walk 'relatively freely' (O'Dell, 2010: 38). Thus during this time, the cure-seeker was more or less the passive recipient of cures that were under medical control, and had to endure treatments that often required them to exercise bodily restraint and discipline. In this way, water cures could be positioned in relation to the evolving associations between health and morality during the 19th century (Verouden and Meijman, 2010).

### Blue space as a medium for health promotion

As medical science progressed however, belief in the effectiveness of the various water cures began to wane, as early as towards the end of the 19th century. There was little compelling evidence for the healing virtues of water, and this was coupled with a growing awareness of the health risks of contaminated water. The treatment of contaminated water to combat water-borne diseases such as cholera, typhoid and malaria, and the consecutive introduction of a common water supply that was clean and safe to drink contributed to a cultural reformulation of the diverse varieties of 'waters' with their own unique chemical properties to the singular conception of water as a 'monolithic substance' containing varying degrees of impurities (Hamlin, 2000).

There is a decisive way in which the idea of the 'water cure' as something to be consumed under specialist medical guidance in dedicated settings where the 'patient' was separated from everyday life has shifted to being associated much more strongly with the everyday through the concept of 'lifestyle'. 'Taking the waters' has today been translated into commodification of healthy lifestyles associated with hydration, exercise and nutrition, and can be seen, for example, in the increased popularity of drinking bottled mineral water, which has led the bottled water industry to be the fastest growing in the world (Brei, 2018). The huge increase in bottled water sales in Western countries where the overwhelming majority of the population has access to safe tap water points to the fact that how water is imagined and represented in policy, politics and marketing has far reaching consequences, not least in terms of public anxieties about risk and health (Wilk, 2006).

Curative practices associated with water occupy a relatively marginal place in terms of contemporary mainstream approaches to healthcare, although water plays an important role in hygiene practices within formal medical settings like hospitals and clinics. However, engagements with water that can be related to self-care and everyday wellbeing practices feature prominently in contemporary leisure pursuits. For example, the contemporary demand for 'wellness' – combining desires for health, pleasure and luxury – has been met by the proliferation of spas, Jacuzzis and saunas (Adams, 2015; O'Dell, 2010). Elements of the water cure, such as hydrotherapy and massage, have continued to be taught and practised, although these treatments are now considered at best a part of physiotherapy rather than medicine, and are no longer associated with natural waters (Adams, 2015) but rather take place in spa settings. Spas are primarily environments where people de-stress, pamper and enhance their beauty; health-related practices in such environments can often be filed under the more fashionable expressions of 'self-improvement', aimed at both the inside and outside of the body (Klepp, 2009; Little, 2013), but may also target overall wellbeing or more specific health goals through activities such as floatation, water aerobics and water sound therapies (Verouden and Meijman, 2010). Water is also present in a range of everyday health and wellbeing related practices, such as at-home bathing and pampering, and the drinking of mineral water, as mentioned above. Engagements with natural waters remain important however, as demonstrated by the growing popularity of outdoor swimming (Atkinson, Chapter 12), exemplified by the membership of the UK Outdoor Swimming Society rocketing from just 300 in 2006 to more than 25,000 in 2016 (Landreth, 2017).

Ongoing research now explores the role of water in contemporary health practices and as part of potentially health-promoting leisurescapes. In public health discourses and policy, the natural environment is increasingly considered as a key site for health promotion, mainly through physical activity (Ashbullby et al., 2013). A wide range of national and local initiatives that promote the use of (predominantly green) outdoor spaces to benefit the health and wellbeing of adults and children can be seen across Europe, the Americas and Australasia (Ashbullby et al., 2013: 138). There is a growing body of scientific evidence across positivistic research arenas that extols the value of green space for healthy living and wellbeing (Lee et al., 2015), and addressing the so-called twin epidemics of obesity and mental illness is no longer the reserve of the health professions but increasingly has become an interdisciplinary research agenda to be tackled by planning, design and the environmental sciences, connecting human and ecosystem health (Tzoulas et al., 2007). Interest in the specific value of water as part of such healthy living environments is still marginal, but has been on the increase, especially in the urban planning and policy domains (de Bell et al., 2017; Gascon et al., 2017; Völker and Kistemann, 2011) as well as ecosystem-based management (Depledge and Bird, 2009; Papathanasopoulou et al., 2016). 'Ecosystems services' is a key framework driving these new forms of valuation, emphasising the costs and benefits of blue environments, aiming to quantify health benefits as part of cost-benefit analyses (e.g. Papathanasopoulou et al., 2016). Within this framework, there is little recognition given to lived meanings of places, or ability to include alternative narratives of value. For instance, what is it about engagements with water, and blue space experiences more broadly, that can contribute to enhanced feelings of wellbeing and health? While it is not possible to prove cause and effect, there is a host of meta-studies and reports that have established a link between spending time in natural environments and subjective wellbeing (Gascon et al., 2017). But these studies tell us very little about the experience of spending time in nature, and what it is about immersion in natural environments that make some people feel better, physically, mentally, emotionally and spiritually. As Humberstone (2015: 63) comments, 'Narratives and tales of being in nature tell of an extraordinary richness to peoples' engagements with places, landscapes and seascapes'. This richness of experience is best captured by qualitative methodologies, which are sensitive to the powerfully emotional and sensory nature of such experiences. Here, the therapeutic landscape concept, with a focus on the cultural, historical and individual factors influencing people's encounters with blue spaces, and its emphasis on the experiential, may open up opportunities to integrate a differently measured 'valuation' of more intangible aspects that promote health and wellbeing in blue landscapes (Bell et al., 2018). The 'therapeutic' value of engagements with water, or blue spaces more broadly, is understood within the therapeutic landscape literature to arise from a dynamic and relational assemblage of experiential, embodied and emotional geographies, which may support the maintenance or promotion of health and wellbeing for different individuals and groups, at different times (Bell et al., 2018), as well as over time (Foley, 2017). Water, thus, is not best understood as a 'curative agent' but rather as part of broader enabling place-assemblages, 'capturing a particular *quality of experience* – an accretion of feelings, capacities, opportunities and interactions – even as it designates a particular moment of enhanced health or wellbeing' (Duff, 2011: 149). The health outcomes of such experiences are unpredictable and in many cases not measurable, and as Foley (2017: 45) comments, these blue space experiences can be active, through physical activity, but can also be passive, in the sense of reflective and imaginative experiences of place (Völker and Kistemann, 2011).

## Blue experiencescapes

Empirical accounts of enabling blue spaces open up understandings of how such spaces affect the experience of place and wellbeing for different groups of people. The literature focusing on experiential approaches to place and health can address the complex questions about how informal therapeutic landscapes may be constituted through mundane and everyday practices, as well as those experiences that rely on a perceptual separation between the restorative setting and everyday life. Paying attention to the diversity of experience within the constitution and cultivation of therapeutic geographies leads to a more comprehensive understanding of how health is produced through engagements with different types of water environments, be it spa retreats or urban canal walks. Grounding research in lived experience can further give us insight into the therapeutic geographies of different social groups. For example, Coleman and Kearns (2015) and Finlay and colleagues (2015) have analysed how different blue settings can both contribute towards wellbeing and pose challenges (e.g. isolation, lack of services) for older adults. Therapeutic engagements with blue spaces need not include immersion in the water; Finlay and colleagues (2015) found that everyday contact with blue spaces (e.g. fountains, ponds, beaches, rivers, lakes and the ocean), which involved physical activity such as walking, but also more passive experiences such as viewing pleasant scenes or listening to birds, had an impact on the sense of mental wellbeing for the older people in their study. Whereas both blue and green spaces have been found to offer opportunities for physical activity and social interactions that play an important part in reports of enhanced physical and social wellbeing across the life-course, Finlay and colleagues (2015) found that contact with blue spaces in particular was experienced as relaxing and stress-reducing, contributing to feelings of renewal, restoration and rejuvenation, as well as affording a sense of spiritual connectedness with loved ones. Blue spaces were found to offer 'multisensory enjoyment, including the sounds of moving water, tranquil surroundings [...] and opportunities in warmer weather to physically experience the water (e.g. dipping one's fingers or feet in the water)' (Finlay et al., 2015: 105). Much of the richness of therapeutic place-engagements, as Humberstone (2015) points out, is expressed through the affective and sensory registers of experience; thus, experiential accounts have included an emphasis on the embodied 'sensescapes' of being in or near water as conducive to feelings of enhanced wellbeing (Foley, 2017; Humberstone, 2015; Straughan, 2012). As illustrated in the section above, the bulk of studies that combine the natural environment and 'healthy living' have focused around a concern with (the promotion of) physical activity. In contrast, however, within the therapeutic landscape literature there have been a growing number of accounts that place affective and emotional dimensions at the centre of their analyses, shifting the focus from a concern with sedentary lifestyles and obesity to an interest in both physical and mental wellbeing and health (Foley, 2017). Examples of such sensory experiences include seeing and hearing the waves, the feeling of wind, the smells of the sea, the sensations of weightlessness and being supported by the water. Here, authors point to the importance of understanding the senses as intertwined and working together, and playing a crucial role in the constitution of feelings and emotions. For example, Straughan (2012) explored the importance of the sense of 'touch' (the 'haptic' system) in the experience of scuba diving. The haptic system registers the contact between the body and the environment through externally felt pressures on the skin, temperature and pain, and the internal sensations provided by nerve endings in the muscles, somatic receptors, hair cells and bodily fluids in the canals of the vestibular (inner ear). These different sensory inputs work together through processes known as 'kinaesthesia', referring to the awareness of the body's position in space through the sense of movement felt in the muscles, tendons and joints; 'it is no single sense but works as a synergetic conjunction or nexus of visceral sensation and exterior perception' (Paterson, 2009: 4). For divers, writes Straughan (2012: 21), it is the physical sensations of the textures of the water that pave way for the emotional responses that her participants described as meditative, giving rise to a sense of tranquillity.

The humanities and social sciences have seen something of an 'affective turn' in the last ten years or so (Ticineto Clough and Halley, 2007), with its most prominent effect being the focus on the lived body and its capacities, feelings and emotions. Affect is a complex term, which, alongside non-representational theory, emotional geographies and experiential perspectives more generally, has been important in recent work on therapeutic landscapes. Affect may be understood as a response on a bodily level which is beyond conscious thought and language (Thrift, 2008), and which facilitates our meaningful engagement with the surrounding environment and each other (Humberstone, 2015). This embodied physicality is important, because it highlights, as Strang (2005) argued, that the meaning of water is not limited to beliefs and ideas, but is also communicated through its very materiality. As Evers (cited in Humberstone, 2015: 64) remarks in relation to his research into embodied surfing masculinities, 'We feel our body and then we experience emotion. Affects are what we feel at the bodily level. Both emotion and affects rely on one another to organize and register our experience with the world'. The final point is important, as it clarifies one prominent line of argument about the relationship between affects and emotions which regards them (although differentiated) to be part of relational and continuous encounters between person and world (Bondi and Davidson, 2011; Spinney, 2015). Most theorisations of affect start from the work of philosopher Baruch Spinoza (or Benedict in the anglicised form), and his treatise *Ethics*. In one of his most often-cited quotations about affect, Spinoza maintained that 'No one has yet determined what the body can do' (de Spinoza, 1994: 155, E3.P2S). The understanding that we can take from this is that the individual body and its affects (and capacity in turn to affect) is an ongoing composition that unfolds within a specific situated context, what can be described as the *'this-ness* of a world and a body' (Gregg and Seigworth, 2010: 3), or what followers of Duns Scotus (or more recently Deleuze and Guattari) term 'haecceity'. It is this ongoing contextual anchoring that makes affect so central to experiential approaches to understanding how health is produced in place. That place is a relational achievement is now something of a mainstay of geographical approaches to how places are made, and made meaningful.

## Towards the diversity of the therapeutic blue

Paying attention to lived, affective and embodied experiences means that research on therapeutic geographies can become more sensitive to the diversity of experiences, capacities, positionalities and power relations of different bodies within these spaces. As Thomas (2015) has argued, some bodies still feel 'out of place' in supposedly enabling blue and green spaces, despite the overwhelming 'evidence' and rhetoric extolling their therapeutic potential. The varied reasons for this include moral tropes around 'healthy lifestyles' and body weight, which can lead many to feel self-consciously open to judgement (Thomas, 2015). As Thomas (2015: 188) points out, despite prevailing positive experiences of the health-related potential of green and blue spaces, awareness is needed of the wider sociopolitical environment, including popular and policy discourse which may make certain bodies experience such spaces as far from restorative. Work in fat studies over the last decade or so has certainly opened up the critical analysis of fat as 'a fluid subject position relative to social norms' (Cooper, 2010: 1021), and within geography and planning accounts have brought to light the exclusionary geographies created by policy efforts to 'design out fatness' (e.g. Evans et al., 2012), by making public spaces both physically and psychologically uncomfortable for those whose bodies do not 'fit', in both a physical and symbolic sense. Though this relates to public spaces and leisure spaces more broadly, blue spaces such as beaches or swimming pools may present particularly stressful environments for some people. The awareness (or worry) about the judging gaze of others when donning the revealing or skin-tight outfits associated with swimming or water sports can present a particular barrier for certain people to experience engagements with water as restorative, and transform 'therapeutic blue space' into the opposite – landscapes of anxiety. Although immersion in water at the same time has the potential to make bodies feel light and mobile in ways, they do not on land. It should thus not be assumed that the physical and mental health benefits of green and blue spaces are universally accessible. Body size is not the only barrier; of course, Byrne and Wolch (2009) have shown that city parks are racialised spaces, which work to exclude some park users. The use of green and blue leisure space is also highly classed, risking the exclusion of those from lower socio-economic groups who do not feel included in fashionable leisure lifestyles (Thomas, 2015). Many outdoor leisure spaces are also disabling for those with a physical or mental disability. These exclusionary geographies have led to concerns that access to so-called 'therapeutic landscapes' remains a 'privilege' of dominant groups within society (Bell et al., 2018; Conradson, 2014). Attention to the experiential infers an understanding of the 'therapeutic' encounter as necessarily contingent on a number of factors to do with how the material, social, symbolic and individual come together in particular contexts. Understandings of therapeutic landscapes that are grounded in an attention to sensory experience, affect and emotion might equip researchers to better engage with 'corporeal contingency and embodied difference' (Bell et al., 2018: 125). An example of the few but growing accounts of diverse experiences of therapeutic landscapes includes a set of Dutch studies of the experiential value of accessible water-sport holidays for people with physical and intellectual disabilities (van den Nieuwenhuizen et al., 2017; Wagenaar and Vaandrager, 2018). When it comes to populations that are in receipt of formalised interventions, such as medical treatment and rehabilitation programmes for either physical or mental health diagnoses, there is often found to be a lack of integration between medically driven care and the social aspects of care that overlap with everyday life (Banks et al., 2016), so there is much scope for accessible forms of water-based activities, for instance, to bridge this gap. Wagenaar and Vaandrager (2018) found that a water-sports weekend provided a transition between a rehabilitation trajectory at a rehabilitation centre and everyday life. They observed that through outdoor experiential learning the participants gained new insights into their bodily capacities, which provided physical and mental resources that could be carried across to other areas of daily life.

There is still a need to explore the potential therapeutic effect of physical immersions in blue space for people with physical and/or intellectual disabilities. As Bell and colleagues (2018) have argued, there needs to be more attention within the therapeutic landscape literature to efforts to increase the accessibility of health-related geographies. Such efforts could further connect the therapeutic landscape literature with the wider literature on geographies of disability, and informal spaces of care.

## Concluding remarks

Paying attention to embodied sensory experiences in blue space is to acknowledge that the impact that such engagements may have on wellbeing and health are relational, intersubjective and situated in time and space. Thus, there is both a broad impetus and scope to investigate blue space as part of the myriad of layered meanings, spaces and activities of everyday life that can potentially make a difference for health and wellbeing. Just like in relation to green space, a number of different agendas converge around the valorisation, promotion and exploitation of blue space. The WHO emphasis on health as more than just the absence of disease and impairment has been reflected in the broad 'wellbeing agenda' that has taken shape across Western countries over the past decade or so. The therapeutic landscape approach is particularly well suited to explore the multitude of spaces where wellbeing, as a complex emergent process, may be realised and promoted, as such researchers working in this area are also well placed to contribute policy-relevant knowledge that is sensitive to differential experiences and needs when it comes to the pursuit of health and wellbeing.

There are multiple lived realities playing out in places that may be described as therapeutic blue spaces, and the promotion and commercialising of such places as sites where people can practise health-enhancing activities, like swimming or water sports, or just enjoying the environment of the coast, is also caught up with efforts to address a host of socio-economic or environmental issues these places are facing. Thus, the value of blue space for health needs to be understood relationally within a broader context where policies at different levels, involving public health, but also local reinvestment, redevelopment, place promotion, UNESCO designations and so on, stand in relation to the lived meanings of these places, and the bottom-up practices that create different kinds of place-values. Understanding some of the different stakes involved in the valorisation and promotion of blue space for health – alongside the multiple discourses and practices that produce different kinds of health-seeking behaviour – is important to clarify the ways social, spatial and political factors may promote or impede health-promoting practices and experiences within particular places, for different people.

# References

Adams, J. (2015). *Healing with Water: English Spas and the Water Cure, 1840–1960.* Manchester, Manchester University Press.

Andrews, G.J. and Holmes, D. (2007). Gay bathhouses: Transgressions of health in therapeutic places. In, A. Williams (ed), *Therapeutic Landscapes.* Aldershot, Ashgate, 221–232.

Ashbullby, K.J., Pahl, S., White, M.P. and Webley, P. (2013). The beach as a setting for families' health promotion: A qualitative study with parents and children living in coastal regions in Southwest England. *Health and Place, 23,* 138–147.

Banks, L., Stroud, J. and Doughty, K. (2016). Community treatment orders: Exploring the paradox of personalization under compulsion. *Health and Social Care in the Community, 24* (6), 657–798.

Bell, S., Foley, R., Houghton, F., Maddrell, A. and Williams, A. (2018). From therapeutic landscapes to healthy spaces, places, and practices: A scoping review. *Social Science & Medicine, 196,* 123–130.

Bell, S., Phoenix, C., Lovell, R. and Wheeler, B. (2015). Seeking everyday wellbeing: The coast as a therapeutic landscape. *Social Science & Medicine, 142,* 56–67.

Bondi, L., and Davidson, J. (2011). Lost in translation. *Transactions of the Institute of British Geographers, 36* (4), 595–598.

Brei, V.A. (2018). How is a bottled water market created? *Wiley Interdisciplinary Reviews: Water, 5* (1), 1–14.

Brown, M., and Humberstone, B. eds. (2015). *Seascapes: Shaped by the Sea.* Farnham, Ashgate.

Brown, K.M. (2016). The haptic pleasures of ground-feel: The role of textured terrain in motivating regular exercise. *Health Place, 46,* 307–314.

Byrne, J. and Wolch, J. (2009). Nature, race and parks: Past research and future directions for geographic research. *Progress in Human Geography, 33* (6), 743–765.

Charlier, R.H. and Chaineux, M.C.P. (2009). The healing sea: A sustainable coastal ocean resource: Thalassotherapy. *Journal of Coastal Research, 25,* 838–856.

Coleman, T. and Kearns, R. (2015). The role of blue spaces in experiencing place, aging and wellbeing: Insights from Waiheke Island, New Zealand, *Health & Place, 35,* 206–217.

Conradson, D. (2014). Health and well-being. In, P. Cloke, Crang, P. and Goodwin, M. (eds), *Introducing Human Geographies* (3rd ed.). London, Routledge, 599–612.

Cooper, C. (2010). Fat studies: Mapping the field. *Sociology Compass, 4* (12), 1020–1034.

de Bell, S., Graham, H., Jarvis, S. and White, P. (2017). The importance of nature in mediating social and psychological benefits associated with visits to freshwater blue space. *Landscape and Urban Planning, 167,* 118–127.

de Spinoza, B. and Curley, E. trans. (1994). *A Spinoza Reader: The Ethics and Other Works.* Princeton, NJ, Princeton University Press.

Depledge, M. and Bird, W. (2009). The blue gym: Health and wellbeing from our coasts. *Marine Pollution Bulletin, 58* (7), 947–948.

Duff, C. (2011). Networks, resources and agencies: On the character and production of enabling places. *Health & Place, 17,* 149–156.

Evans, B., Crookes, L., and Coafee, J. (2012). Obesity/fatness and the city: Critical urban geographies. *Geography Compass*, 6 (2), 100–110.

Finlay, J. (2018). 'Walk like a penguin': Older Minnesotans' experiences of (non) therapeutic white space. *Social Science & Medicine,* 198, 77–84.

Finlay, J., Franke, T., McKay, H. and Sims-Gould, J. (2015). Therapeutic landscapes and wellbeing in later life: Impacts of blue and green spaces for older adults. *Health & Place,* 34, 97–106.

Foley, R. (2010). *Healing Waters: Therapeutic Landscapes in Historic and Contemporary Ireland.* Ashgate: Aldershot.

Foley, R. (2011). Performing health in place: The holy well as a therapeutic assemblage. *Health & Place,* 17 (2), 470–479.

Foley, R. (2014). The Roman–Irish bath: Medical/health history as therapeutic assemblage. *Social Science & Medicine,* 106, 10–19.

Foley, R. (2015). Swimming in Ireland: Immersions in therapeutic blue space. *Health & Place*, 35, 218–225.

Foley, R. (2017). Swimming as an accretive practice in healthy blue space. *Emotion, Space & Society,* 22, 43–51.

Foley, R. and Kistemann, T. (2015). Blue space geographies: Enabling health in place. *Health & Place,* 35, 157–165.

Gascon, M., Zijlema, W., Vert, C., White, M. and Nieuwenhuijsen, M. (2017). Outdoor blue spaces, human health and well-being: A systematic review of quantitative studies. *International Journal of Hygiene and Environmental Health,* 220, 1207–1221.

Gesler, W. (1992). Therapeutic landscapes: Medical issues in light of the new cultural geography. *Social Science & Medicine,* 34 (7), 735–746.

Gray, F. (2006). *Designing the Seaside: Architecture, Society and Nature.* London, Reaktion Books.

Gregg, M. and Seigworth, G. (2010). *The Affect Theory Reader.* Durham, NC, London, Duke University Press.

Hamlin, C. (2000). "Waters" or "water"? Master narratives in water history and their implications for contemporary water policy. *Water Policy,* 2, 313–325.

Humberstone, B. (2015). Embodiment, nature and wellbeing: More than the senses? In, M. Robertson, Lawrence, R. and Heath, G. (eds), *Experiencing the Outdoors: Enhanced Strategies for Wellbeing.* Rotterdam, Boston, Taipei, Sense Publishers, 61–72.

Kearns, R., Collins, D. and Conradson, D. (2014). A healthy island blue space: From space of detention to site of sanctuary. *Health & Place,* 30, 107–115.

Klepp, I.G. (2009). Does beauty come from within: Beauty and well-being in Norwegian spas. *Medische Anthropologie,* 21(1), 39–51.

Kruizinga, P. (2016). Health tourism and health promotion at the coast. In, M. Smith and Puczkó, L. (eds), *The Routledge Handbook of Health Tourism.* London, Routledge, 386–398.

Landreth, J. (2017). Brrr! The joys of cold water swimming. *The Telegraph,* 13 February 2017. Available at: https://www.telegraph.co.uk/health-fitness/body/brrr-joys-cold-water-swimming/, last accessed June 6, 2017.

Lee, A., Jordan, H. and Horsley, J. (2015). Value of urban green spaces in promoting healthy living and wellbeing: Prospects for planning. *Risk Management and Health Policy,* 8, 131–137.

Lengen, C. (2015). The effects of colours, shapes and boundaries of landscapes on perception, emotion and mentalising processes promoting health and well-being. *Health & Place,* 35, 166–177.

Linton, J. (2010). *What is Water? The History of a Modern Abstraction.* Vancouver, UBC Press.

Little, J. (2013). Pampering, wellbeing, and women's bodies in the therapeutic spaces of the spa. *Social & Cultural Geography,* 14, 41–58.

Lykke Syse, K.V. and Oestigaard, T. (eds) (2010). *Perceptions of Water in Britain from Early Modern Times to the Present: An Introduction.* Bergen, Uni Global.

O'Dell, T. (2010). *Spas: The Cultural Economy of Hospitality, Magic and the Senses.* Lund, Nordic Academic Press.

Papathanasopoulou, E., White, M., Hattam, C., Lannin, A., Harvey, A. and Spencer, A. (2016). Valuing the health benefits of physical activities in the marine environment and their importance for marine spatial planning. *Marine Policy,* 63, 144–152.

Porter, R. (ed) (1990). *The Medical History of Waters and Spas: Introduction.* London, Wellcome Institute for the History of Medicine.

Paterson, M. (2009). *The Senses of Touch: Haptics, Affects and Technologies.* Oxford, Berg Publishers.

Pitt, H. (2014). Therapeutic experiences of community gardens: Putting flow in its places. *Health & Place,* 27, 84–91.

Spinney, J. (2006). A place of sense: A kinaesthetic ethnography of cyclists on Mont Ventoux. *Environment and Planning D: Society and Space,* 24, 709–732.

Spinney, J. (2015). Close encounters? Mobile methods, (post)phenomenology and affect. *Cultural Geography,* 22 (2), 231–246.

Strang, V. (2005). Common senses: Water, sensory experience and the generation of meaning. *Journal of Material Culture,* 10 (1), 92–120.

Straughan, E. (2012). Touched by water: The body in scuba diving. *Emotion, Space and Society,* 5, 19–26.

Thomas, F. (2015). The role of natural environments within women's everyday health and wellbeing in Copenhagen, Denmark. *Health & Place,* 35, 187–195.

Thrift, N. (2008). *Non-representational Theory: Space/Politics/Affect.* London, Routledge.

Ticineto Clough, P. and Halley, J. (eds) (2007). *The Affective Turn: Theorizing the Social.* Durham, NC, Duke University Press.

Tzoulas, K., Korpela, K., Venn, S., Yli-Pelkonen, V., Kazmierczak, A., Niemela, J. and James, P. (2007). Promoting ecosystem and human health in urban areas using green infrastructure: A literature review. *Landscape and Urban Planning,* 81 (3), 167–178.

van den Nieuwenhuizen, J., de Jonge, F.H., Berends, E., van der Horst, M., de Jong, N., Jonkers, I., Liem, W., van Nimwegen, J., Agro, P., Peters, S., van der Pol, E., Schmitz, P., Sluimer, N. and Wagenaar, M. (2017). *De Meerwaarde van watersport-activiteiten voor mensen met een beperking: Eindreportage.* Wageningen, Wageningen University & Research, Wetenschapswinkel (Wetenschapswinkel rapport 337). Available at: http://library.wur.nl/WebQuery/wurpubs/529288, last accessed June 2, 2017.

van Tubergen, A. and van der Linden, S. (2002). A brief history of spa therapy. *Annals of the Rheumatic Diseases,* 61, 273–275.

Verouden, N.W. and Meijman, F.J. (2010). Water, health and the body: The tide, undercurrent and surge of meanings. *Water History,* 2, 19–33.

Völker, S. and Kistemann, T. (2011). The impact of blue space on human health and well-being – Salutogenetic health effects of inland surface waters: A review. *International Journal of Hygiene and Environmental Health,* 214 (6), 449–460.

Völker, S. and Kistemann, T. (2015). Developing the urban blue: Comparative health responses to blue and green urban open spaces in Germany. *Health & Place,* 35, 196–205.

Wagenaar, M. and Vaandrager, L. (2018). Experiencing a watersport holiday as part of a rehabilitation trajectory: Identifying the salutogenic mechanisms. In, I. Azara, Micholopolou, E., Niccolini, F., Derrick Taff, B. and Clarke, A. (eds), *Tourism, Health and Wellbeing in Protected Areas.* Wallingford, CABI, 167–177.

Wilk, R. (2006). Bottled water: The pure commodity in the age of branding. *Journal of Consumer Culture,* 6 (3), 303–325.

# 7  *Dúchas*

## Being and belonging on the borderlands of surfing, senses and self

*Easkey Britton*

### Introduction

Within global surfing culture, a process of reclamation is underway to understand surfing as a boundary-crossing, fluid, sensual and reciprocal experience (Anderson, 2012; Britton, 2018; Brown and Humberstone, 2016; Evers, 2006; Lisahunter, 2012, 2016; Olive, 2013, 2015; Wheaton, 2013). There is growing interest in how surfing could be used as medium to challenge social norms and inequalities (Britton, 2018; Comer, 2010; Olive et al., 2015; Roy and Caudwell, 2014) as well as the tensions and power dynamics inherent in the cultural politics of surfing as it becomes increasingly formalised and commercialised (Britton et al., 2018; Wheaton, 2013). However, very little study has focused on *how* we learn to *be* in the natural environment (Humberstone, 2013) and the relational processes and place connections that might be driving how we experience ourselves in nature, especially water (Brown and Humberstone, 2015). As scholars of health geographies within 'blue space' have written, our experience of immersion in water influences our sense of wellbeing and self-connection (see Chapter 4, in this book; Britton, 2018; Foley, 2015; Strandvad, 2018; Straughan, 2012; Throsby, 2013, 2015). This chapter explores how our experience of the sea, through sharing the experience of surfing, can facilitate a greater sense of *relational* wellbeing (White, 2017). White (2017) argues that wellbeing emerges through the dynamic interplay of personal, societal and environmental processes. This relational understanding of wellbeing emphasises the influence of our relationship with the natural environment on our wellbeing and views wellbeing as a process of 'interactions among the material, organic and emotional dynamics of place' (Atkinson, 2013: 138). For the purpose of this chapter, I wish to link this relational understanding of wellbeing with the Irish/Gaelic concept of *dúchas*. The meaning translates broadly as natural and cultural heritage, birthright, native place, natural state and homeland (Grady, 1999); or a sense of belonging and place connection. *Dúchas* allows for a rethinking of our heritage and sense of self as a more relational process, as it includes spiritual and environmental associations and emphasises how stories shape our world (McBride, 2018). Throughout this chapter, I will draw upon *dúchas* as lens to understand an auto-ethnographic experience of surfing and its impact on relational wellbeing. In the following section, I briefly outline the cultural connection to the sea in the context of Ireland.

The sea is both a place of loss and healing. In Ireland, the narrative of the sea is traditionally one that has centred on the sea as a place of loss associated with tragedy, poverty and penance. There has been a collective forgetting of the sea since famine times (1800s) with the departure of what became known as the 'coffin ships', the term given to ships that carried Irish immigrants to America escaping the Great Famine. This turning our backs on the sea has led to a cultural disconnect (de Courcy Ireland, 1992) that has persisted into the 21st century. It is perhaps most evident in the lack of political will with regard to the sustainable management of our marine resources (Gilmartin, 2009). However, the very characteristics associated with loss – wild, unpredictable, stormy – are being reimagined following the launch of the Wild Atlantic Way in 2014, a highly successful marketing campaign by Failte Ireland that is changing how our coasts are valued once more (Failte Ireland, 2017). Other emerging initiatives point to a remembering and re-awakening of our ocean heritage such as the recently established national Ocean Literacy Network, the nation's annual Sea Fest, Love Your Coast photo competition, wild swimming (Foley, 2015), and surfing and other water sports becoming the economic mainstay in many coastal communities. This points to a new narrative of the sea in Ireland, and one that fits within a wider global shift, e.g. blue gym initiative (Depledge and Bird, 2009), blue health (Foley and Kistemann, 2015; Grellier et al., 2017) and ocean optimism (Kelsey, 2016).

In many ways, in our globalised society we are more (technologically) connected than we have ever been. However, there are increasing media reports claiming that we are the loneliest society that has ever been. We live in an age of growing disconnect. Otto Scharmer, co-founder of the Presenting Institute, outlines three great divides of our age (Scharmer and Kaufner, 2013): (i) ecological, the 'border' we've created between ourselves and the natural environment, e.g. ocean plastic, rising sea surface temperatures and an age of mass extinctions; (ii) social, between ourselves and others, e.g. resurgent nationalism, intensified border controls and a refugee crisis with more displaced people in the world than ever before; and (iii) spiritual/psychological, between our inner and outer selves, e.g. rocketing mental health issues and suicide rates, rise in 'industrial health epidemics' (e.g. heart disease, type 2 diabetes and obesity). In this age of disconnect, as White states, 'It is vital that narratives of wellbeing generate an expanded and socially inclusive vision and practice' (White, 2017: 129). Recent 'Blue Mind' research spearheaded by marine biologist Wallace J. Nichols, combining neuroscience and psychology, highlights how water environments might be one way to achieve this. He argues water environments can be especially powerful for helping us bridge these divides by better connecting with ourselves, each other and nature (Nichols, 2014).

In this chapter, I illustrate using several auto-ethnographic accounts of surfing how the notion of 'borders' (divides or boundaries between our sense of self and 'other') can be challenged. I use the concept of borderland consciousness, a 'sociosensual awareness' of our ecological reality (Bernstein, 2005; Totton, 2011: 74), to explore my own position at the intersection of surfing, self, senses and place. What that means for my identity as both a surfer, scholar and someone who lives in a border county of Ireland is also considered. I present two accounts from embodied experiences that illustrate an entanglement with the senses, woven into and through the dynamic and fluid environment of the sea. First, big-wave surfing in Ireland, and second, a surf-related programme with women in Papua New Guinea (PNG) that illustrates how embodied experiences in this liminal space could challenge notions of duality and otherness. These accounts draw on Pink's definition of 'sensory ethnography' where we as researchers become co-creators, *'sharing and empathising with other people's experiences and actions'* (as cited in Lisahunter and Emerald, 2016: 33). Taking this reflexive auto-ethnographic approach (Humberstone, 2011; Rinehart, 2005), I discuss what these experiences mean for how we understand ourselves, each other and our environment. *Dúchas* is used to help provide a more storied and relational understanding of place and sea connection. The influence of history and power (over) is also revealed through the lens of *'dúchas'*. In conclusion, I discuss what this could mean for how we might apply a more bodily way of knowing that emphasises relational wellbeing and allows for a rethinking of our nature-culture connection through a language of the senses.

## Borderlands of surfing, senses and self

My name 'Easkey' (or *Iascaigh*) has its origins in the Gaelic for 'fish'. It's also the name of a world-class surf break on the west coast of Ireland, one of my parents' favourite waves. It does make me wonder at the power of naming (Rollins, 1985). Traditional naming rituals were an important rite of passage historically, such as *he'e nalu* or surfing lineage in Hawaiian culture, and understanding the influence that can have on our decisions and our story. Rachel Carson (as cited in Alaimo, 2012) follows these aquatic roots back into the sea, highlighting the evolutionary connection between the sea and all living creatures:

> When they went ashore the animals that took up a land life carried with them a part of the sea in their bodies, a heritage which they passed on to their children and which even today links each land animal with its origin in the ancient sea.
>
> (p. 482)

Stacy Alaimo adds how, 'The sea surges through the bodies of all terrestrial animals, including humans – in our blood, skeletons, and cellular protoplasm'

(Alaimo, 2012: 482). We ourselves are bodies of water yet we are not separate from water bodies. However, on the borderlands of nature-culture a strong dichotomy persists between the two. This is particularly evident in blue space research (Depledge et al., 2013; White et al., 2010, 2014). There is a persistent 'othering' of nature as separate from culture, where humans are removed from nature. This is evidenced in the 'back-to-nature' discourse in the literature (Louv, 2012), and an overemphasis on outdoor experiences and adventure activities in 'wilderness', removed from everyday life (Comer, 2016; Mitten, 1994). There is a tendency to neglect race, ethnicity or sexuality from these studies and nature therapy is largely aligned with Western values to the exclusion of other value systems and meanings of nature (e.g. the practice of *shinrin yoku* or 'forest bathing' in Japan or the *atua matua*, a holistic Maori health framework (Warbrick et al., 2016)). These dominant approaches dismiss the role of power, power that is formed at 'the intersection of privilege in terms of race/class/gender/species/sexuality' (Neimanis, 2014: 29). Furthermore, the 'politics of water' (Neimanis, 2014) and contestation of coastal spaces (Britton et al., 2018) are often ignored. For example, the persistent pay gap between female and male athletes in surfing (Franklin, 2009) (and almost every other sport), the devaluation of women's shore and care work in fisheries (Britton, 2013; Coulthard and Britton, 2015), and the ongoing privatisation and selling out of natural resource rights in Ireland, the most recent casualty being traditional seaweed harvesting (Siggins, 2016). Cultural heritage scholar Liam Campbell (2016) explains what this means in the context of a postcolonial Ireland, for how we might understand the loss of self which results from colonisation:

> It is widely accepted that Gaelic peoples have been through a colonial experience. [...] The desire to control the land and the resources of other peoples and countries in order to enrich or safeguard the core is a defining characteristic of imperialism ... [T]oday the nature of interventions and resources are different, ranging from natural gas to wind, but analogous processes continue to operate.

Some scholars argue that this disconnection from land is intergenerational. Evidence from epigenetics (Skinner et al., 2010) reveals how the impact of environmental stress (such as the loss of land) can lead to the transgenerational inheritance of trauma and related abnormalities and disease, or what Skinner (2014: 153) refers to as 'ancestral stress'. Furthermore, this 'leaves an imprint on the epigenome (...) carried into future generations with devastating consequences' (Walters et al., cited in Warbrick et al., 2016: 400). This can lead to a loss of connection to our landscapes, environments and the storied knowledge they hold (Johnson, 2010). The next section provides an auto-ethnographic account of how the sea shapes a sense of being and belonging.

## The sea as a place of human experience

Neimanis' (2013: 23) concept of how we 'reside both within and as part of a fragile global hydrocommons,' seems to mirror my upbringing and childhood on the north-west of Ireland. I live on the borderlands – the geographical and political borders of my origin in the remote border county of Donegal; the border between my skin and the sea; the border between land and water. Borderlands, it has been written, can be places of wounding and bleeding, and yet the scars signify healing. My early childhood experiences were influenced by the end of the so-called 'Troubles'. This was a period of intense sectarian violence and armed paramilitary conflict for decades in Northern Ireland and its border regions. My parents lived through some of the height of it, especially my mother who was a student at University of Ulster in Northern Ireland. I remember as a child the border checkpoints and sometimes having the car my mother was driving stripped and searched, including my surfboards, on the way to surfing contests in the North Antrim coast. Yet, during the height of the conflict, surfing had an ability to transcend Nationalism and identity politics and bring together a mix of surfers from 'North' and 'South', from Protestant and Catholic communities for surfing events. Surfing contests were (and still are) an all-island affair and Antrim was a regular stop on the Irish surfing circuit. The Rossnowlagh Surfing Intercounties, founded by my grandfather in 1968, included the six counties of Northern Ireland. These stories have strongly influenced my sense of place.

My own story has been greatly shaped not only by my place connections but also by my 'blue heritage' – an intergenerational connection with the sea and surfing that has been passed on like a genetic code through stories. I can trace this back to my grandmother on my father's side, an enterprising woman who ran a hotel business on the Atlantic coast of Donegal, the remotest, most rural and socially deprived region of Ireland, established in the 1960s. At the same time, she was promoting tourism in Ireland on a visit to California in the 1960s at the height of the emergence of contemporary surf culture and remarked many decades later to me how she thought the waves of Rossnowlagh (the beach in front of her hotel) were far superior to the ones she was looking at from her hotel in Malibu. It inspired her to invest in some of the first surfboards in Ireland. Being the mother of five sons, it was not long before they got their hands on the surfboards and began trying to learn what to do with the boards in the cold surf, learning what they could from travelling, foreign surfers and experimenting with the first wetsuit designs. On reflection, the stories that we carry within us in terms of our heritage can have a profound influence on our sense of identity and purpose. I was born into this pioneering surfing family and I still surf regularly with my father who is in his sixties. I cannot remember a time in my life 'before surfing'. As a child, we'd go on family road trips and camp next to the breaking surf. My younger sister and I curled up between our parents in the back of the van, the waves crashing on the reef. I remember staying up late to listen for a rise in the sound of the waves that signalled a shift of tide or the arrival of a new swell. I learned about the reef, the swell and tides from time spent in rock pools, observing what the sea left behind when the tide ebbed and watching them fill in as my father timed his surf for the flooding tide. To borrow from Sara White, the sea taught me that "the natural world has its own processes, flows and constraints, rhythms and tipping points which respond to human action but are also significantly beyond human control" (White, 2017: 131).

I'm keenly aware of the multiple identities or roles I play as surfer, teacher, scholar, activist and artist and how we increasingly find ourselves having to take on new and different roles in a rapidly changing world. So how then do we navigate this shifting terrain and not lose our connection or sense of belonging? To address this question, I return to the concept of *dúchas,* rather than place or landscape. There is no word in the Irish language for 'landscape', because landscape is not viewed as something separate from self. *Dúchas* is associated with a sense of belonging or place-anchoring, similar to Gruenewald's description of cultural experience as 'the "ecology" of the diverse relationships that take place within and between places' (2008: 137). Surfing, for me, is a powerful exchange of energies between salt water and body surface. It is like an embodied mindfulness practice, the act of doing something with your whole being (or a *body mindfulness,* see Bleakley, 2016). It also helps me understand the meaning of connection or connectedness:

> Surfing demands a letting go of expectations and requires our total presence in the moment so that we begin to take notice of our surroundings, to be a little awed and humbled through direct contact with nature. It's as if the salt water literally dissolves the stress, worries or fears that we carry on land, the barriers we put up within ourselves and the world outside, and we are no longer separate but a part of it all.
>
> (Britton, 2013, speaking at TEDx Dublin)

This notion of presence is strongly woven into more ancient Gaelic and Celtic world views. In the song of Amergin (Gregory, 1904), the chief poet of the Milesians (another 'coloniser' group to arrive in Ireland), there is a line, '*I am because everything else is. I am everything and everything is in me*'. A belief that is common across many indigenous cultures and ancient spiritual traditions.

There is a beautiful Gaelic phrase, *fighte fuaighte,* which Irish philosopher John O'Donohue (1998: X) described as 'woven into and through each other'. By this, he meant how in Celtic world view there was no separation of the senses, self, nature or indeed between the eternal and living worlds. The act of surfing could be framed in a similar way, bringing the environment 'out there' back into our way of being, thinking and acting (Neimanis et al., 2015). These notions of interconnectedness, entanglement and permeability and the fluid nature of water bring on a whole new meaning when we consider the notion of borderlands. Being 'immersed' in the sea, our skin and bodies permeable, we are

> invaded by terrestrial, human consumerism, revealing the swirling natural-cultural agencies, the connection between ordinary terrestrial life and ocean ecologies, and the uneven distribution of harm.
>
> (Alaimo, 2012: 488)

Similarly, anthropologist Tim Ingold (2011: 86) argues for the need to regain a sensory awareness of landscape and describes how we are *'not limited by the skin. It, too, leaks'*. Our desire to be cleansed, to wash away the stress and worries carried on land, has become tainted in an ocean environment that also carries our terrestrial waste. I've directly experienced the consequences of surfing in water contaminated by untreated sewage overflow at one of Ireland's most popular surfing spots, with no warning for water users of the risks (see also Wheaton, 2007), and the higher risk of exposure to antibiotic resistant bacteria in coastal bathing waters for surfers (Leonard et al., 2015). This further emphasises the dynamic and fluid nature of our relationship with the natural environment and how that can hinder or enable our sense of wellbeing.

## Researcher position – 'the active observer'

> One is continually present as witness to that moment, always moving like the crest of a wave, at which the world is about to disclose itself for what it is.
>
> (Ingold, 2011: 69)

By outlining my own sense of self and place, I hope to account for the location of my knowledge and its limits, while also highlighting what Evers refers to as the 'ethnographer's body' and how the *'surfing ethnographer's body can be a recording of sensual data'* (Evers, 2006: 501). In participatory action research we must consider the question of (mis)representation, that is, the issue of speaking on behalf of others and for the more-than-human world (Malone, 1999). Neimanis (2013: 30–32) argues that *'facilitating opportunities for conversation, rather than representation, need to be actively generated'* and a recognition that *'truth is partial and unfolding, and perpetually open to contestation'*. With regard to the 'voice' of the more-than-human world, it is believed in Gaelic traditions that the environment itself has its own language and that this environmental language is the root of human language (Liam Campbell, pers. comm. 9.9.16) (see also MacFarlane, 2015; Sapir, 1912). This is well represented in the origins of Gaelic-Irish place-names that offer rich descriptions of local environmental forces and habitats (Joyce, 1883). This notion of communicating with nature exists in many indigenous belief systems. Maori scholars and health practitioners, Warbrick et al. (2016) give the example of how the sound of a crashing wave can forewarn the danger of entering the ocean, as well as a health programme where regional physical activities are defined by weather patterns, lunar cycles and tidal movements.

In the next section, I provide auto-ethnographic accounts from Ireland and PNG of surfing. The cross-cutting essence is how we might become like water (i.e. hydrophilia unbounded). That is, an entangled, enmeshed and embodied understanding of who we are through surfing. First, I present an account of a big-wave surfing session at a place called 'Mullaghmore' on the north-west coast of Ireland that emphasises the notion of 'edgework'. That is, being on the edge of or 'at the boundary of control and chaos' (Humberstone, 2013: 565; Lyng, 1990). In the second section, there are a number of approaches being initiated and/or adopted by a small community of surfers at Tupira Surf Club (TSC) in PNG aimed at creating a platform for voice in local decision-making and in the wider global surf community, in particular for female members.

The methodological process I adopted during these visits to PNG (for a period of two to three weeks at time in January/February 2015 and 2016) included the sharing and constructing of stories grounded in lived experiences through a mix of active participation, participant observation, informal interviews and photo-voice elicitation. This mixed method approach allowed me to better understand how these young women who surf not only see themselves but experience themselves – who they are, how their bodies move and feel in these watery environments. My role was to observe and support the collaborative potential of these informal, community-based activities and to document the potential for a new way of learning and doing surfing. Karen Malone (1999) took a similar methodological approach in her activist research. She emphasised the need to develop a *'participatory relationship between researcher and researched'*, to support the participants and their own self-documenting, *'as a means for self-reflection and as a source of empowerment'* (Malone, 1999: 233).

## Embodied experiences of surfing

### *I'm not an island, I'm a body of water – Bedouine*

This section illustrates how surfing is fostering new relationships with women, the sea and society and how water environments can be especially powerful for helping us better connect with ourselves, each other and nature. I acknowledge the challenge and limitation of translating the sensuous learning experiences into written text. Lisahunter and Emerald (2016: 37–39) discuss the complexity of this and provide a framework for how to collect, record, capture and communicate the senses and sense experiences, which have been useful for my own interpretation. Similar to others who have written from a sensuous geography perspective about their experiences immersed near (Wylie, 2005), in (Thorsby, 2013, 2015), on (see Chapter 4, this book; Humberstone, 2011) or under the sea (Straughan, 2012; Strandvad, 2018), I include forms of narrative and descriptive writing as a creative and critical means of drawing out a more bodily experience and sensuous knowing. It could be argued that each of these accounts of the sea and surf represents not only a relational place but also a place of sanctuary, or what is known as *tearman* in Gaelic. *Tearman* was the name given to specific sites or places on or near borderlands that offered sanctuary from judgement or persecution in Ireland. However, some of the barriers and challenges are also considered, such as the possible negative implications that can come when 'attempting to shift the balance of power' (Malone, 1999: 239).

## *Ireland: move like water*

The surfed wave, Anderson (2012) argues, is a relational place where a process of convergence emerges. The lines of movement of the surfer and the wave are woven together until there is no separation between self, surfboard and wave. Instead, each component – human, weather process, swell, sea, wind, tide, lunar cycle, rocky reef – is *'contributing its continuing story to this constellation'* (Anderson, 2012: 574). Those moments when we are most deeply immersed in our environments, through an embodied experience, is known as the flow state (Csikszentmihalyi, 1997). My own experience of big-wave 'tow' surfing in winter on the west coast of Ireland seeks to illustrate how the concept of the wave as a relational place and wave-riding as a process of convergence may be experienced in practice (i.e. surfing as a form of *dúchas*) (Figure 7.1). To give context, 'tow surfing' or 'tow-in surfing' is the term used to refer to a highly specialised form of extreme surfing in

*Figure 7.1* Easkey Britton big-wave surfing at Mullaghmore.
Source: Christian McLeod.

big waves, usually greater than 20 feet in height, often much bigger. A jet ski, with a tow rope extended out to the surfer, is used to assist the surfer in catching the wave in situations where the swells are consider to move too fast to be caught by 'paddle power' alone (i.e. the strength of the surfer's arms paddling herself onto a wave). The kind of swells that generate breaking waves of this magnitude are very rare and might only occur a handful of times in a given season and only at a few remote locations around the world, known as 'big-wave spots'. Therefore, the surfers who ride these waves require in-depth local knowledge for how each of these 'spots' works, including the necessary conditions (wind direction, swell size, tide, etc.), and intense physical and mental training and preparation to ensure safety in a high-risk environment. This is a process that can take years and is continuous.

As a woman who has been recognised as a 'big-wave surfer', I must briefly discuss the gendered aspect of big-wave surfing. It is a contested space for women in what has traditionally been viewed as a male domain, despite women surfing alongside men in big-wave spots since contemporary surf culture began (History of Women's Surfing, n.a.). In surf media, big-wave surfing has typically been portrayed as the 'hero' overcoming the 'monster', a hegemonic, hyper-masculine portrayal of a 'dragon-slaying' pursuit that is at odds with what is in reality a highly intimate act of both commitment and surrender. The actual experience is often one of coming together, if you will, of both masculine (doing) and feminine (being) energies with self-reported feelings of oneness and bliss. This lack of representation and recognition of feminine identities and female surfers in the sport came to a head in 2016 following the controversial exclusion of women from a big-wave surfing contest called 'Titans of Mavericks' in California. Contest director, Jeff Clark, was reported in the media saying, 'It's not a gender thing – it's a performance thing. We have a really good understanding of who's performing the best, who is pushing the limits, who is going to new levels of performance. Women just aren't there yet' (CBS News, 2016). This was a blatant dismissal of female big-wave surfers who have been surfing Mavericks since Dr Sarah Gerhardt became the first woman to surf it in 1999. In response, a committed group of female big-wave surfers established the Committee for Equity in Women's Surfing (CEWS) to lobby for greater equality. As a result, the California Coastal Commission (the state agency charged with overseeing public use of the coast) demanded change and women were included in the event. This is situated within a wider discourse on the political position of visual narratives of 'female surfers' and the call for new narratives in surfing or 'different ways of ways of knowing surfing, remembering who and what constitutes/ed surfing' (Lisahunter, 2016: 319).

Unlike traditional sports where there is a defined goal to achieve, the purpose of surfing is the act itself. What then is the drive to enter into these high-risk situations? I have wondered if my attraction to 'edgework', a term used by sociologist Stephen Lyng to describe 'voluntary risk-taking activities that are about exploring edges' (Strandvad, 2018: 57), is somehow shaped by my *dúchas*. That is, my sense of place and belonging at the 'edge', in a border region and the influence of my blue heritage. As the author of *Testosterone Rex*, Cordelia Fine wrote, 'offspring don't just inherit genes. They also inherit an entire "developmental system": an ecological legacy of place, physical environment, and structures; and a social legacy of parents, relatives, peers, and others who also provide important and reliable inputs as the animal grows and learns' (Fine, 2017: 96). The risks we choose to take are shaped by our psychological drive and the affordances around that, regardless of sex (Fine, 2017). What follows is an auto-ethnographic account of 'edgework' at a big-wave spot Mullaghmore in the north-west of Ireland:

It was December 21st 2013, winter solstice, close to sunset. The light was soft and golden breaking through the clouds and the wind died while the swell peaked just before dark. The biggest waves I'd ever been out in. I can't explain why but it's the moment I felt like I finally belonged out there. I was right where I was meant to be because it reinforced why I did it – to be so fully present in such intense nature, to be part of that movement of energy, to experience the joy of that. To go to the edge, that borderland where sea meets rock, and find my aliveness in this temporary place of practice and action.

I can see the wave coming, building momentum and I move to position myself to match its speed so I can meet it at just the right instant. At first I turn away from the wave, towards the shore, my grip tightening on the tow-rope, my vision narrowing. Feeling the pull of the tension on the rope, shifting my feet slightly in the straps through the thickness of my neoprene boots. I can no longer see the wave so much as feel it. The texture and surface of the water change as the wave of energy begins to shape the skin of the sea from beneath the surface until it connects with the rocky reef under the sea and then it is fully able to form, breaking free. It comes fully to life, towering over me. At its apex, at the height of its speed, I catch it and ride it. Again, I only know this when I sense a shift in speed, an acceleration and sudden steepening of the wave face. Most of all, it is the sound. Through my hooded wetsuit I can hear a deep rumbling intensify into a roar like all the world has become sound. It darkens as the wave opens up, breaking over my head, blocking out the weak winter sunlight. It is a moment of total commitment when all that build-up and mental and physical noise and buzz stop. Nothing exists but total awareness in the moment. But the act of wave riding can also be so fleeting. My awareness shifts, my vision or focus widening to take in the other surfers in the line-up once I am blasted by the force of the ride into the safety of the channel. The wave and all its intensity can be gone just as suddenly as it came, and with it I let out a loud exhalation.

It is through a language of the senses that the wave is understood and expressed by the surfer. Not only does the surfer come to understand herself through a more bodily and sensuous knowing but the wave is also expressed and given meaning through the surfer. It is these moments of becoming and belonging for both the wave and the surfer that could be understood as *dúchas*. The act of wave-riding is a coming together of two bodies of water, and a rejecting of body-(mind)-nature dualism. Or as Anderson (2012: 576) explains, '*it is only through the act of surfing that the wave exists*'. This experience of non-duality is not exclusive to surfing but can be found in other 'extreme sports', and is echoed in Sara Malou Strandvad's (2018) auto-ethnographic free-diving account. The other aspect of this sensory experience is the cold water. Even old scars can be felt again as the skin tightens against the cold. Again, maybe that's the nature of my upbringing somewhere so isolated – that strong sense of place, and extreme Irish weather that seems to shape my identity both physically and psychologically. There are also those days when it doesn't all align. They can be an equally powerful embodiment, if not more so. The outcome of these intense surfing sessions is often determined in what I experience as the 'space in between'. That edgespace or borderland between the decision and the action. What happens in the moments before I enter the sea, how present I am, and the environmental and social conditions (how big a crowd is gathered to watch, the emotional state and comments of fellow surfers, etc.) will determine the outcome of what happens in the sea. On one such day, although I committed to paddling out and deeply desired to catch a wave, I didn't commit to throwing myself over the ledge of breaking water and taking off on one of the waves that day. I had many competing voices in my head as to why I didn't. I feel it came down to not having the right motivation. In particular, not being able to let go of the expectation of others, and maybe the expectations I'd placed on myself as a woman in the world of big-wave surfing and how I might be judged. I felt too much focus was on the need to somehow perform well and live up to expectations. Instead of feelings of courage and self-worth, feelings of self-doubt, shame and regret can surface. Especially in an era where every wave ridden is documented and glorified instantaneously on camera. The spaces in between the waves go undocumented.

### Papua New Guinea: following the solowara

In this section, I outline the current seascape of PNG, the development of surfing and a more detailed account of the gendered experience surfing spaces from the perspective of local female surfers and how this relates to a more bodily way of knowing that emphasises relational wellbeing. The seascape of PNG is a highly contested space facing multiple stressors and an arena of competing (largely foreign) interests and pressures on natural resource use/extraction. The resilience of ecosystems has been further eroded by growing vulnerability to climate change with an increase in the frequency of major droughts, cyclones and flooding (Matthews et al., 2012). PNG also falls drastically low on the Gender Inequality Index (a component of the Human Development Index), ranking 140 out of 146 countries in 2012. According to a report on gender, livelihood security and coastal communities, violence against women is particularly prevalent,

> *Polygamy is practiced and early marriage (as young as 13 and 14) is common as well as violent 'witch hunts' and the practice of 'bride price', which contributes to the prevalence of domestic violence. On-going political instability and chronic law and order issues further threaten women's safety. The outcomes are a high rate of HIV/AIDS, high rates of early pregnancy and one of the highest global rates of maternal mortality.*
>
> *(Matthews et al., 2012: 76)*

It is within this dynamic and complex seascape that surfing is emerging as an example of community-based resource management and sustainable tourism (Abel and O'Brien, 2015; O'Brien and Ponting, 2013). The Surfing Association of PNG (SAPNG) implements a formal, community-based Surf Management Plan, an approach that emphasises sustainable tourism development with direct benefits for local 'host communities' in designated surf zones (O'Brien and Ponting, 2013). Although incredibly niche, surfing or some variation of wave-riding on timber boards (known traditionally as *palangi*) has long been practised in PNG, perhaps for millennia (Ponting, 2004). O'Brien and Ponting (2013: 162) have written how some northern coastal communities still practise precolonial surf rituals, 'designed to cause the sea to rise up and provide good quality surf' (Ponting, 2004).

In the following section, I draw on recent experiences surfing with female surfers in the northern province of Madang in a surf zone managed by TSC (established in 2008) and the surrounding communities. TSC, a former logging site, is an example of how surfing is replacing harmful extraction industries as a viable alternative. I discuss the impact of an initiative known as the 'Pink Nose surfboards' (PNS), as well as a recent return to more traditional surfboard-making using timber. The PNS initiative was launched by SAPNG in 2015 to use surfing as a tool aimed at giving voice to previously taboo issues and to help break down the silence around a culture of gender inequality and violence against women (Abel and O'Brien, 2015; Britton, 2015). Despite the tradition of surfing on *palangi*, surfers in PNG are primarily dependent on donations of contemporary surfboards (which they call 'fibreglass boards') from foreign charities and tourists. SAPNG manage the collection and donation of surfboards to surf zones around the country. In 2015, SAPNG implemented a policy to paint half of all the noses (front end of surfboards) pink. These were designated for female surfers only. In addition, a language of equality was included in SAPNG policy and introduced in community meetings and public events. The general consensus among local surfers at TSC in 2016 (a year after the PNS launch) was that the surfboards had given greater recognition and respect to female surfers. However, uptake by new participants, especially younger girls, was slow. Training days like the *Meris* (*Women's*) *Surf Days* served to reinforce and give life to the Pink Nose philosophy and bring together the neighbouring communities as well as providing a platform for some of the younger surfers to develop their own leadership style and skills by sharing knowledge and experience of the sea and surfing with newcomers (Figure 7.2). It also privileged the position of women in the line-up/community/village, for that surf day at least. A self-proclaimed 'surfing mum' from Tavulte village explained how she thought the PNS made a difference, 'In the past, we go around arguing and having conflicts about what boards to take. When the boards are painted pink, that's the difference – the girls have a chance to go out and surf'.

There is no precise word for 'waves' in *Tokpisin* (the national language of PNG). Instead a 'language of the sea' has developed across *Tokples* (local languages) and *Tokpisin* that give different meanings to the experience of the sea or *solowara* (meaning literally, 'saltwater'). Although there does not

*Figure 7.2  Meri* (Women's) Surf Day, Papua New Guinea, 2016.
Source: Nicki Wynn.

seem to be an equivalent to *dúchas,* the words used to describe the experience of surfing are rich in their embodiment of a sociosensual experience. For example, *'mipela go biainim solowara'* translated literally means 'we go follow the sea/waves'. A local surfer and patron of TSC explained how he prefers to use *'biainim solowara'* to *'karim solowara'* (to carry the waves/sea) because, 'we follow the waves. The waves carry us, we don't carry the waves' (Justice Nicholas Kirriwom, pers. comm. 11.6.18).

> When I started body surfing at 6 years and on, I spoke no other language except Mauwaake (tokples). When we as kids go to body surf at Sarar inlet, we say 'yii iifera ookmik.' Yii stands for 'we', iifera the sea/waves, and ookmik to follow. We are following the sea/waves.
>
> (Justice Nicholas Kirriwom, patron TSC, June 2018)

At TSC, I observed the opening of spaces that might allow for new ways of doing surfing. Tupira's timber surfboard project created an alternative way of approaching surfing. A collaborative initiative between local and Australian surfboard shapers seeks to breathe contemporary life into an ancient culture by honouring old ways while creating something new. The hand-shaping of timber boards fosters a sense of autonomy and self-actualisation, and unlike fibreglass boards, timber boards don't create a dependency on imported and/or donated products from 'wealthier' (Western) surfing nations. The timber board lends itself to an exceptional and intimate wave knowledge and an ability to *feel* the waves through the characteristics of the various types of wood. It's perhaps no surprise then that my own more intimate experience of the sea happened after I'd taken a break from fibreglass boards to ride a timber 'belly' board just for fun. The following excerpts from my field journal illustrate a process of convergence where the thresholds between self, board and nature are blurred (Anderson, 2012: 570), and how our experience of self and how we feel is 'mediated through our bodies' (Shilling, 1993: 22):

> This afternoon I touched upon it for a fleeting moment. I felt this sensation of ease and fluidity. My body relaxed, I felt the rise of the wave and let it lift me, I trusted its momentum using less strokes into the wave because I somehow sensed or felt when the right moment would be to drop in. Letting my board and body fall into the wave with the lip, feeling the speed from the drop-in fuel my body into the bottom turn, knowing I didn't have to rush at the lip of the wave but instead letting the moment come, standing taller and looser, flowing with and following the curve of the wave face. Loose but strong, the board firm under my feet, maximising the force of the wave, meeting with its kinetic energy at all the right moments with a new kind of empathy. It felt more effortless, fluid, graceful and playful. Like there was no try.
>
> (27 January 2016)

For me, it was an experience of how to be in the world more fully – noticing how we are feeling in the moment and how that can be expressed through our bodies. Another word sometimes used by local surfers that could capture the sense of embodiment is *waswas*, which means to literally be washed or cleansed by the sea. A process of nature connection where the boundaries between body, board and water are blurred. In this way, surfing becomes a collaboration between what the surfer wants to do and what the ocean is offering. And this dynamic is changing with each micro-movement, moment by moment.

> My last session at Tupira is spent on timber boards with talented local surfer, 17 year old Ruthie. A little bit of PNG magic under our bellies as we fly across the reef feeling like both a novice again and like this is the most natural thing in the world. Ruthie is learning to shape her own timber boards. Her surfing abilities and regular presence in the surf, perhaps aided by the ready access to surfboards for female surfers as a result of the Pink Nose surfboard initiative and encouraged by the recognition of her equal status in the line-up, is translating into the shaping bay. A space traditionally male-dominated the world-over.
>
> (Field notes, 27 January 2016)

The impact of these 'empowerment' initiatives beyond Tupira is harder to gauge. Responses to the actual impact seem mixed. Symbolically and conceptually it is a simple yet powerful tool, and it is encouraging an attitudinal shift in certain domains of life. According to Justice Nicholas Kirriwom, it seems especially impactful in terms of raising awareness and providing a stronger platform for women to contribute to decision-making at the village meetings. Although it improves accessibility and sense of ownership for some female surfers, it doesn't improve for all women. Painting the nose of surfboards pink will not lead to equality or liberation for women without greater structural changes in other spheres of life and a wider network of support services allowing women greater time freedom and autonomy to be able surf. As one female surfer commented, women always have '101 tasks or chores they must do before they can think about surfing' (30 January 2016). Therefore, this change often requires a renegotiation of roles and identities beyond the surf. To this end, TSC are investing profits from the community-owned and -operated surf camp, in education, training courses in tourism and hospitality for young women, provision of healthcare services and, most recently, traditional timber surfboard shaping workshops. In this way, surfing is the tool or medium through which change can be enacted.

## Conclusion

> The coming ashore is often the hardest part. I felt more 'lost at sea' on land, unsure which point to focus on back on the too-stable terrain.

It was like I had to let the particles settle into a new pattern after the days of elemental disruption. For now I let it be, inviting the stillness in, mindfully feeling the ground beneath my feet before moving forward again.

(Field notes, February 2016)

What I have described in this chapter are only snapshots of my surf-related experiences in Ireland and PNG. They seek to highlight, in Humberstone's words (2011: 495), how the 'shared experience and values of people in nature-based sport and leisure in different spaces across the globe can provide a realisation of common values'. This is evidenced by the shared sense of water connection among the women from across the global North-South divide. Writing this chapter has been part of my own uncovering of place-based knowledge as fluid and dynamic, understanding place as a sense of belonging or a relational process of becoming known as *dúchas*. As Johnson (2010: 6) states, this recovery of place connection 'means recognising the importance of particular places within our lives', especially the wave as a relational place (Anderson, 2012: 570). This self-reflective and sensory auto-ethnographic approach represents an attempt to reclaim and reconstruct a sense of place and identity. It seeks to 'move us usefully beyond the political problems of (b) order control derived from a sedentary version of place' (Anderson, 2012: 584), and towards a more dynamic notion of place as a convergence of flows, connections and relationships (Anderson, 2012; Braun, 2004). It is both the flow (of relationships) and the constraints (of power structures), as well as place (including the forces within the natural environment beyond human control), which are important for how we understand a more relational way of being (White, 2017).

These accounts reveal a more bodily way of knowing that allows for a rethinking of our nature-culture connection through a language of the senses. I have illustrated a variety of forms and mediums where embodied knowledge and practice are valued and legitimised such as the timber surfboards. In turn, this has led to the development of alternative discourses (e.g. relational wellbeing, cultural heritage, identities, sensory ethnography, restorative benefits of surfing, blue space). In essence, each account highlighted how we experience our bodies of water in water bodies. Perhaps by rethinking the ways in which we learn and do surfing and why we do it we can come to a greater relational understanding of ourselves, each other and our environment. However, there are still many barriers and challenges to be considered. As much as there are opportunities, there are also negative implications that can come when the balance of power shifts. Although experienced as places of sanctuary or *tearman*, between worlds or at the edge of existing worlds – the reclaiming of the sea and surfing as a relational place for minority groups, especially women, can threaten existing patriarchal power structures and hyper-masculine societies. With this threat, there is potential for gendered conflict and/or disregard for these new modes of being and doing in the water by those who hold traditional positions of power (Britton et al., 2018; Comer, 2010). There is a need to continue to recognise and give voice to 'those who contest surfing through their non-normative surfing bodies' (Lisahunter, 2016: 319).

Ultimately these accounts can help us understand how we relate to each other and our natural environment, which means, in addition to a relational understanding of place, we need a relational understanding of who 'we' are. There are many ways of *being* a surfer and many different ways of *doing* surfing. We need to celebrate the diversity of the lines we can draw and how we can move through the water and waves. As local surfer from PNG and mother to up-and-coming young surfers (including her eight-year-old daughter Mary) explained, it is about how we might embody the unique qualities that can be found at the heart of surfing, leading to a greater sense of self-connection and awareness:

> One more big thing about surfing – Surfing really is a work-out for your brain and how you focus. It's about how you see what you are doing and you make change for yourself, and keep improving.
>
> (Female surfer, Tupira, February 2016)

## References

Abel, A.C. and O'Brien, D. (2015). Negotiating communities: Sustainable cultural surf tourism. In, G. Borne and Ponting, J. (eds), *Sustainable Stoke: Transitions to Sustainability in the Surfing World*. Plymouth, University of Plymouth Press, 154–165.

Alaimo, S. (2012). States of suspension: Trans-corporeality at sea. *ISLE: Interdisciplinary Studies in Literature and Environment,* 19 (3), 476–493.

Anderson, J. (2012). Relational places: The surfed wave as assemblage and convergence. *Environment and Planning D: Society and Space,* 30 (4), 570–587.

Atkinson, S. (2013). Beyond components of wellbeing: The effects of relational and situated assemblage. *Topoi,* 32 (2), 137–144.

Bernstein, J. (2005). *Living in the Borderland*. London, Routledge.

Bleakley, S. (2016). *Mindfulness and Surfing: Reflections for Saltwater Souls*. Leaping Hare Press.

Braun, B. (2004). Nature and culture: On the career of a false problem. In, J. Duncan, Johnson, N. and Schein, R. (eds), A *Companion to Cultural Geography*. Malden, MA, Blackwell, 151–179.

Britton, E. (2013). *The Pink Nose Revolution*. Maptia, published online Available at: https://maptia.com/easkey/stories/the-pink-nose-revolution

Britton, E. (2015). Just add surf: The power of surfing as a medium to challenge and transform gender inequalities. In, G. Borne and Ponting, J. (eds), *Sustainable Stoke: Transitions to Sustainability in the Surfing World*. Plymouth, University of Plymouth Press, 118–127.

Britton, E. (2018). 'Be like water': Reflections on strategies developing cross-cultural programs for women, surfing and social good. In, M. Mansfield, Caudwell, J., Watson, R.

and Wheaton, B. (eds), *The Palgrave Handbook of Feminism and Sport, Leisure and Physical Education*. London, Palgrave Macmillan, 793–807.

Britton, E., Olive, R. and Wheaton, B. (2018). Surfers and Leisure: 'Freedom' to surf? Contested spaces on the coast. In, M. Brown and Peters, K. (eds), *Living with the Sea* (pp. 161–180). London, Routledge.

Brown, M. and Humberstone, B. (eds) (2015). *Seascapes: Shaped by the Sea*. Farnham, Ashgate Publishing, Ltd.

Brown, M. and Humberstone, B. (2016). Seascapes: Surfing the sea as pedagogy of self. In, *Seascapes: Shaped by the Sea*. London, Routledge, 57–70.

Campbell, L. (2016, July 27). *Belonging and Exile - Duchas and Dearaiocht. First Nations Ireland*. Accessed on 1 January 2019, https://firstnationsireland.wordpress.com/2016/07/27/belonging-and-exile-duchas-and-deariocht/

CBS News. (2016, February 12). *Wave of controversy hits California surfing competition*. Accessed on 1 January 2019, https://www.cbsnews.com/news/titans-of-mavericks-surfing-competition-california-no-women-surfers/

Comer, K. (2010). *Surfer Girls in the New World Order*. Durham, NC: Duke University Press.

Comer, K. (2016). "We're blacksurfing": Public history and liberation politics in white wash. *Journal of American Ethnic History, 35* (2), 68–78.

Coulthard, S. and Britton, E. (2015). Waving or drowning: An exploration of adaptive strategies amongst fishing households and implications for wellbeing outcomes. *Sociologia Ruralis, 55* (3), 275–290.

Csikszentmihalyi, M. (1997). *Flow and the Psychology of Discovery and Invention*. New York, Harper Perennial.

de Courcy Ireland, J. (1992). *Ireland's Maritime Heritage*. Dublin, An Post.

Depledge, M.H., Harvey, A.J., Brownlee, C., Frost, M., Moore, M.N. and Fleming, L.E. (2013). Changing views of the interconnections between the oceans and human health in Europe. *Microbial Ecology, 65* (4), 852–859.

Depledge, M.H. and Bird, W.J. (2009). The blue gym: Health and wellbeing from our coasts. *Marine Pollution Bulleting, 58* (7), 947–948.

Evers, C. (2006). How to surf. *Journal of Sport and Social Issues, 30* (3), 229–243.

Failte Ireland. (2017, May 9). *New Wild Atlantic Way TV Ad encourages Irish people to get away from it all*. Accessed on 9 January 2019, http://www.failteireland.ie/Utility/News-Library/New-Wild-Atlantic-Way-TV-Ad-Encourages-Irish-Peopl.aspx

Fine, C. (2017). *Testosterone Rex: Unmaking the Myths of Our Gendered Minds*. London, Icon Books.

Foley, R. (2015). Swimming in Ireland: Immersions in therapeutic blue space. *Health & Place, 35*, 218–225.

Foley, R. and Kistemann, T. (2015). Blue space geographies: Enabling health in place. *Health & Place, 35*, 157–165.

Franklin, R. (2009). *Recognition for Female Surfers: Riding a Wave of Sponsorship? Educational Planet Shapers: Researching, Hypothesizing, Dreaming the Future*. Australia, Post Pressed.

Gilmartin, M. (2009). Border thinking: Rossport, Shell and the political geographies of a gas pipeline. *Political Geography, 28* (5), 274–282.

Grady, A. (1999). Natural and cultural heritage – From conflict to symbiosis. In, Nature as Heritage: From Awareness to Action: Proceedings, Strasbourg, 3–4 June 1999. Council of Europe publishing, *Environmental Encounters, 47*, 27–33.

Gregory, L.A. (1904) [1976]. *Gods and Fighting Men: The Story of the Tuatha De Danaan and of the Fianna of Ireland.* Gerrards Cross, Colin Smythe.

Grellier, J., White, M.P., Albin, M., et al. (2017). Blue health: A study programme protocol for mapping and quantifying the potential benefits to public health and well-being from Europe's blue spaces. *BMJ Open,* 7: e016188. doi:10.1136/bmjopen-2017-016188

Gruenewald, D.A. (2008). The best of both worlds: A critical pedagogy of place. *Environmental Education Research,* 14 (3), 308–324.

Humberstone, B. (2011). Embodiment and social and environmental action in nature-based sport: Spiritual spaces. *Leisure Studies,* 30 (4), 495–512.

Humberstone, B. (2013). Adventurous activities, embodiment and nature: Spiritual, sensual and sustainable? Embodying environmental justice. *Motriz: Revista de Educação Física,* 19 (3), 565–571.

Ingold, T. (2011). *Being alive: Essays on movement, knowledge and description.* London, Routledge.

Johnson, J.T. (2010). Indigeneity's challenges to the settler state: Decentring the 'imperial binary'. In, T. Banivanua Mar and Edmonds, P. (eds), *Making Settler Colonial Space.* London, Palgrave Macmillan, 273–294.

Joyce, W. (1883). *The Origin and History of Irish Names of Places* (Vol. 2). Dublin, MH, Gill.

Kelsey, E. (2016). The rise of ocean optimism. *Hakai Magazine,* June 2016 Issue.

Leonard, A.F., Zhang, L., Balfour, A.J., Garside, R. and Gaze, W.H. (2015). Human recreational exposure to antibiotic resistant bacteria in coastal bathing waters. *Environment International,* 82, 92–100.

Lisahunter. (2012). Surfing life: Surface, substructure and the commodification of the sublime. *Sport, Education and Society,* 17 (3), 439–442.

Lisahunter. (2016). Becoming visible: Visual narratives of 'female' as a political position in surfing: The history, perpetuation, and disruption of patriocolonial pedagogies? In, H. Thorpe and Olive, R. (eds), *Women in Action Sport Cultures* (Global Culture and Sport Series). London, Palgrave Macmillan, 319–347.

Lisahunter and Emerald, E. (2016). Sensory narratives: Capturing embodiment in narratives of movement, sport, leisure and health. *Sport, Education and Society,* 21 (1), 28–46.

Louv, R. (2012). *The Nature Principle: Reconnecting with Life in a Virtual Age.* Chapel Hill, NC, Algonquin Books.

Lyng, S. (1990). Edgework: A social psychological analysis of voluntary risk taking. *American Journal of Sociology,* 95 (4), 851–886.

Macfarlane, R. (2015). *Landmarks.* Harmondsworth, Penguin.

Matthews, E., Bechtel, J., Britton, Morrison, K. and McClennen, C. (2012). *A Gender Perspective on Securing Livelihoods and Nutrition in Fish-Dependent Coastal Communities.* Report to the Rockefeller Foundation from Wildlife Conservation Society, Bronx, NY.

Malone, K. (1999). Environmental education researchers as environmental activists. *Environmental Education Research,* 5 (2), 163–177.

McBride, L. (2018). *The Triskele Process: A handbook for change.* Ebook, Kindle Edition. Accessed at: https://www.amazon.co.uk/Triskele-Process-Handbook-Change-ebook/dp/B07LCWQ2RZ

Mitten, D. (1994). Ethical considerations in adventure therapy: A feminist critique. *Women & Therapy,* 15 (3–4), 55–84.

Neimanis, A. (2013). Feminist subjectivity, watered. *Feminist Review,* 103 (1), 23–41.

Neimanis, A. (2014). Natural others? On nature, culture. In, M. Evans, Hemmings, C., Henry, M., Johnstone, H., Madhok, S., Plomien, A. and Wearing, S. (eds), *The SAGE Handbook of Feminist Theory.* London, Sage, 26–44.

Neimanis, A., Åsberg, C. and Hedrén, J. (2015). Four problems, four directions for environmental humanities: Toward critical posthumanities for the Anthropocene. *Ethics & the Environment,* 20 (1), 67–97.

Nichols, W.J. (2014). *Blue Mind: The Surprising Science that Shows How being Near, in, on, or Under Water Can Make You Happier, Healthier, more Connected, and Better at What You Do.* London, Hachette.

O'Brien, D. and Ponting, J. (2013). Sustainable surf tourism: A community centered approach in Papua New Guinea. *Journal of Sport Management,* 27 (2), 158–172.

O'Donohue, J. (1998). *Anam Cara: A Book of Celtic Wisdom.* London, Harper Perennial.

Olive, R. (2013). Blurred lines: Women, subjectivities and surfing. *PhD Thesis,* School of Human Movement Studies, The University of Queensland.

Olive, R. (2015). Reframing surfing: Physical culture in online spaces. *Media International Australia, Incorporating Culture and Policy,* 155, 99–107.

Olive, R., McCuaig, L. and Phillips, M.G., (2015). Women's recreational surfing: A patronising experience. *Sport, Education and Society,* 20 (2): 258–276.

Ponting, J. (2004). Wali Kam: An organic surf culture blooms in New Guinea. *The Surfers Journal,* 13 (4), 5–8.

Rinehart, R.E. (2005). "Experiencing" sport management: The use of personal narrative in sport management studies. *Journal of Sport Management,* 19 (4), 497–522.

Rollins, R. (1985). Friel's 'Translations': The ritual of naming. *The Canadian Journal of Irish Studies,* 11 (1), 35–43.

Roy, G. and Caudwell, J. (2014). Women and surfing spaces in Newquay, UK. In, J. Hargreaves and Anderson, E. (eds), *Routledge Handbook of Sport, Gender and Sexuality.* London, Routledge.

Sapir, E. (1912). Language and environment. *American Anthropologist,* 14 (2), 226–242.

Scharmer, C.O. and Kaufer, K. (2013). *Leading from the Emerging Future: From Ego-System to Eco-System Economies.* Oakland, CA, Berrett-Koehler Publishers.

Shilling, C. (1993). *The Body and Social Theory.* London, Sage.

Siggins, L. (2016). Sea change in the west: Deal divides seaweed harvesters. *Irish Times,* June 18th, 2016.

Skinner, M.K., Manikkam, M. and Guerrero-Bosagna, C. (2010). Epigenetic transgenerational actions of environmental factors in disease etiology. *Trends in Endocrinology & Metabolism,* 21 (4), 214–222.

Skinner, M.K. (2014). Environmental stress and epigenetic transgenerational inheritance. *BMC Medicine,* 12 (1), 153.

Strandvad, S.M. (2018). Under water and into yourself: Emotional experiences of freediving contact information. *Emotion, Space and Society,* 27, 52–59.

Straughan, E.R. (2012). Touched by water: The body in scuba diving. *Emotion, Space and Society,* 5 (1), 19–26.

Throsby, K. (2013). 'If I go in like a cranky sea lion, I come out like a smiling dolphin': Marathon swimming and the unexpected pleasures of being a body in water. *Feminist Review,* 103 (1), 5–22.

Throsby, K. (2015). 'You can't be too vain to gain if you want to swim the Channel': Marathon swimming and the construction of heroic fatness. *International Review for the Sociology of Sport,* 50 (7), 769–784.

Totton, N. (2011). *Wild Therapy: Undomesticating Inner and Outer Worlds.* Monmouth, PCCS Books.

Warbrick, I., Dickson, A., Prince, R. and Heke, I. (2016). The biopolitics of Māori biomass: Towards a new epistemology for Māori health in Aotearoa/New Zealand. *Critical Public Health,* 26 (4), 394–404.

Wheaton, B. (2007). Identity, politics, and the beach: Environmental activism in surfers against sewage. *Leisure Studies,* 26 (3), 279–302.

Wheaton, B. (2013). *The Cultural Politics of Lifestyle Sport.* London, Routledge.

White, S.C. (2017). Relational wellbeing: Re-centring the politics of happiness, policy and the self. *Policy & Politics,* 45 (2), 121–136.

White, M.P., Wheeler, B.W., Herbert, S., Alcock, I. and Depledge, M.H. (2014). Coastal proximity and physical activity: Is the coast an under-appreciated public health resource? *Preventive Medicine,* 69, 135–140.

White, M., Smith, A., Humphryes, K., Pahl, S., Snelling, D. and Depledge, M. (2010). Blue space: The importance of water for preference, affect, and restorativeness ratings of natural and built scenes. *Journal of Environmental Psychology,* 30 (4), 482–493.

Wylie, J. (2005). A single day's walking: Narrating self and landscape on the South West Coast Path. *Transactions of the Institute of British Geographers,* 30 (2), 234–247.

# 8 Blue yogic culture

## A case study of Sivananda Yoga Retreat, Paradise Island, Nassau, Bahamas

*Allison Williams, Ashleigh Patterson and Morgan Parnell*

### Introduction: enabling health through therapeutic landscapes

This chapter explores the importance of blue spaces, that is, environments rich in access to water and sky, in the context of spirituality at a yoga ashram and vacation spot in the Bahamas. We will discuss the ways that therapeutic landscapes enable health, how blue spaces do this in particular and the impacts of spirituality in these environments.

Wellbeing is defined as being 'content, healthy, and in a good place in life' (Andrews et al., 2014: 212). There are many varying opinions on how to best ensure the wellbeing of our population, from the promotion of welfare states on the political left to the insistence on 'economic wellbeing' on the right (Andrews et al., 2014). In geographies of health, our focus is on how the concept of place fits into developing 'a state of positive mental and physical health and welfare, attained or obtained in some way by fulfilling personal needs' (Andrews et al., 2014: 213; Kearns, 1993). It is obvious that place has a significant impact on physical, mental and spiritual health (Gesler, 1993; Keniger et al., 2013). This is seen from the context of religious pilgrimage to our day-to-day lives (Williams, 2007). With that said, places do not inherently cause positive or negative health but instead enable differential health outcomes (Andrews et al., 2014; Bell et al. 2018; Duff, 2012; Foley and Kistemann, 2015).

This is especially true of mental health, where restorative natural environments have been shown to relieve stress and anxiety (Duff, 2012; Mitchell, 2013). Duff (2012) wrote that the therapeutic benefits of place likely have more to do with the ways they allow and encourage specific types of social interaction than with the places themselves. This aligns with Williams' (1999) discussion of how subjective experiences of places give them meaning and these experiences include social interaction, essential activities and aesthetic qualities. Groenewegen et al. (2012) also wrote about the importance of social interaction in green spaces and how access to green space builds a sense of community in residential areas.

Natural environments in particular seem to reduce stress and promote faster recovery from illness (Foley and Kistemann, 2015; Groenewegen et al., 2012; Hartig et al., 2014; Mitchell, 2013). Keniger et al. (2013) found that access to green space can mediate the 'individual isolation, lack of social support, interracial conflict and increased incidence of crime and violence' associated with urban spaces. Bell et al. (2015) point out that people will not necessarily use green spaces – or blue spaces – simply because they have access to them. That said, in interviews people tend to identify these natural spaces with benefits for both physical and mental health which may be a motivating factor to use them (Bell et al., 2015). Exercise is also perceived as being easier when performed in natural environments (Gladwell et al., 2013). Many studies have discussed the benefits of nature exposure for young children, especially (Keniger et al., 2013). It is unclear whether it is necessary to have interactions with natural environments frequently and repeatedly over the life-course to see positive health impacts or if a single intensive exposure can have the same beneficial effects (Pearce et al., 2016). It is particularly difficult to assess the long-term health-enabling benefits of spending time in natural environments (Hartig et al., 2014).

Keniger et al. (2013) acknowledge that the extensive literature purporting the health-enabling benefits of spending time in nature may be biased towards environments at higher latitudes where issues such as venomous snake bites and zoonotic diseases are less likely to be encountered. Williams (2007) also mentions that what is therapeutic for one person or group of people may not be therapeutic for everyone, depending on their local and personal contexts. There is also a problem of correlation versus causation in research studying the impacts of natural environments on health, that is, self-selection for studies, and the possibility that healthier people seek out green spaces more often because they are able to be both confounding factors in this research (Groenewegen et al., 2012).

Therapeutic landscapes are made up of four key elements: built environment, natural environment, psychosocial environment and spiritual environment. Within these environments we can measure physical, mental, psychosocial and spiritual health. Very few studies have focused on the spiritual benefits of interacting with nature (Keniger et al., 2013; Williams, 2007). In those studies that do exist, benefits such as 'increased inspiration and feelings of connectedness to broader reality' were identified (Keniger et al., 2013: 926). In the context of yoga culture, Hoyez (2007) writes about the importance of natural elements, cultural elements, emotions and social relations to the growing number of yogic therapeutic landscapes globally. She discusses the globalised production and reproduction of these therapeutic spaces and places made up of both built structures and natural physical elements, such as water and green space, as important to the spiritual practice of yoga (Hoyez, 2007).

## The importance of blue spaces

While there is much literature on the experience of 'green spaces' and their impact on health, there are less data on how we interact with 'blue spaces' on their own (Foley and Kistemann, 2015; Völker and Kistemann, 2011). Blue spaces are health-enabling places where water is at the focal point of human wellbeing (Duff, 2012). Gesler (1992) discussed how many societies have used water as a source of healing. From the classical Greeks and Romans to contemporary seeking of rural landscapes, bodies of water have been seen as having curative and restorative powers (Gesler, 1992; Williams, 1999). In many faiths, water is used in rituals and ceremonies as a method of purification (Völker and Kistemann, 2011).

As time has progressed, the safety of swimming and interacting with water bodies has increased in most places – the advent of lifeguards and vaccinations against waterborne disease are examples of safety measures that have increased access to water activities (Collins and Kearns, 2007). Unlike green spaces, blue space is more accessible to a wide variety of bodies as even those who are disabled in most environments can experience freedom in immersion in water (Foley and Kistemann, 2015).

Völker and Kistemann (2011) found four dimensions of blue space as healing – experienced space, symbolic space, social space and activity space. Each of these plays an important role in the way water enables health. Experienced space focuses on the removal from everyday life, spiritual on 'supernatural' healing powers, social on practices such as shared rituals and activity on things such as exercise in water (Völker and Kistemann, 2011). Collins and Kearns (2007) identify beaches in New Zealand in particular as being therapeutic landscapes because they provide: distance from everyday life, access to the natural environment, the option for either solitude or social interaction, an opportunity to shape identities and opportunities for recreational/athletic endeavours. It follows that these qualities would be true of most beach areas around the world, including the ashram where this current research was conducted.

It is important to recognise that even in 'blue spaces', there are other colours and sensory experiences present. Most natural 'scenescapes' include blues, greens, browns and greys along with scents, tastes and proprioceptory stimulus (Nettleton, 2015; Spinney, 2006; Straughan, 2012), all of which have potential to contribute to health and wellbeing (Lengen, 2015).

Therapeutic landscape theory has undergone a number of iterations since first unveiled by Gesler (1992). Theorizing therapeutic landscapes has moved from a predominately descriptive focus to one that increasingly concentrates on the simultaneously embodied, emotional and experiential aspects of therapeutic spaces (Bell et al., 2018). The recent use of colour, as illustrated in this edited text, is yet another layer of theorisation in the ongoing development of the therapeutic landscape concept. The recognition of such places as healthful to some and harmful to others has also been a growing theme. The changing nature of health tourism is also of interest here, where holistic retreats tend to offer combinations of therapies and counselling, pathways to spiritual development, creative enhancement and many other routes to the reconciliation of body, mind and spirit (Smith and Kelly, 2016).

## Context of the Ashram

This research was conducted at the Sivananda Ashram Yoga Retreat on Paradise Island, Bahamas. The ashram was founded in 1967 and is located on a beach across the bay from Nassau on 4.4 acres of lush tropical gardens. The Paradise Island Sivananda Ashram combines a traditional ashram with a wide variety of workshops, courses, musical performances and special events, all organised by the core staff who live on site (see https://www.sivanandabahamas.org/program-calendar/). The ashram has winding, curved paths through the gardens that lead from the beach villas and tent/accommodation areas, temples, yoga platforms and the various work areas, such as laundry, carpentry and food preparation. It is a 15-minute walk along the beach or through the forest to a large vacation casino resort, putting the sacred in juxtaposition with the profane.

The philosophy of the ashram community is rooted in the teachings of Swami (a Hindu male religious teacher) Sivananda (1887–1963), and Swami Vishnudevananda (1927–1993) founded the International Sivananda yoga Vedanta Centres (http://www.sivananda.org/teachings/swami-vishnudevananda.html). Swami Vishnudevananda is known to have created the first yoga teachers training course, and to synthesise a complex and extensive set of ancient teachings into five principles of hatha yoga (http://www.sivananda.org/teachings/swami-vishnudevananda.html). Swami Vishnudevananda believed that individual peace, gained through yoga and meditation, would translate into international peace worldwide (Williams, 2017). The Yoga Teacher Training Program at the ashram is one month long and encourages participants to take what they've learned and to spread it to their communities at home, especially to marginalised populations (Williams, 2017). A karma (service) yoga programme is also offered; it is three months long and requires five to six hours of labour daily, from food preparation through to gardening and administration. There is unscheduled 'down time' at the ashram, allowing for social interaction between classes, workshops and karma yoga duties (Williams, 2017). Meals are served twice a day and are all vegetarian (Williams, 2017). For those open to it, the ashram offers a space of healing and rejuvenation through the environment and spiritual practices there.

The ashram has increased in popularity in recent years as it began offering yoga teacher training during the summer months as well as promoting the place as a vacation retreat. Each year they run a minimum of nine-month-long yoga teacher training courses in addition to advanced yoga teacher training, specialised yoga teacher training and the karma yoga programme. The ashram site was originally the home of a wealthy family, who passed on the property to the yogic community concerned following the healing of their daughter from drug addiction. She partook in the spiritual practices of yoga, breathing and meditation under the guidance of Swami Vishnudevananda.

## Methods

Following approval from the university ethics office, the first author gained permission from the ashram to schedule a week-long field work session in February of 2017. During this week, two main data collection methods were used: participatory observation, as well as in-depth semi-structured interviews. Snowball sampling allowed seven interviews to be conducted with either permanent staff or regular residents (who visit for extending periods throughout the year) of the ashram. These groups were targeted given their depth of experience with the ashram over an extended amount of time, as well as their experience with thousands of guests. The semi-structured interview schedule (see Box 8.1) was made up of only a few questions, allowing probes to be used throughout. The process of the interviews allowed other place-specific responses to emerge, specifically around the importance of the blue space. The audio-taped interviews averaged one hour and took place in various quiet places throughout the ashram.

Following transcription, the interviews were thematically coded by the authors. For the first three interviews, the authors individually generated lists of keywords from the transcripts then grouped them by themes. Following this manual coding process, which allowed the student co-authors to master thematic coding, the remaining four interviews were coded using Nvivo qualitative analysis software. The authors met weekly to discuss and confirm themes, in addition to checking in regularly through the coding process.

---

### BOX 8.1 INTERVIEW SCHEDULE, SIVANANDA ASHRAM

1 How is the Sivananda Ashram and the associated Ashram activities (i.e. daily meditation, yoga classes) healing: (a) spiritually?, (b) physically?, (c) emotionally? and (d) psychologically?
2 Have you noticed whether there are any sex or gender differences in the use of the ashram and the outcomes experienced therein?

---

## Findings

As outlined in Table 8.1, the participant sample was made up of four women and three men, ranging in age from 35 to 71. The average number of years at the ashram was 25.

*Table 8.1* Participant characteristics, Sivananda Ashram Yoga Retreat

| Participant pseudonym | Country of origin | Sex | Approximate age | Approximate years associated with the ashram |
|---|---|---|---|---|
| Paula | Canadian | Female | 46 | 10 |
| Jack | Canadian | Male | 69 | 50 |
| Ossa | Canadian | Female | 71 | 26 |
| Jamie | Swiss | Male | 35 | 15 |
| Garda | Israel | Female | 42 | 17 |
| Cookie | Canada | Female | 61 | 30 |
| Lockie | Canada | Male | 58 | 26 |

In total four themes were determined with respect to the blue space of the ashram: (1) safe and protected space, (2) the gendered experience, (3) healing energy and (4) time extended.

## Safe and protected sacred blue space

As identified in literature, one of the many health-enabling benefits of spending time in natural environments is that it is separate from everyday life. The ashram amplifies this experience through the expectations of self-less service, yoga philosophy, clean living and eating, and a safe, inclusive environment that is free from violence. Ossa said:

> So it's a harm-free environment, I would say – to the body, to the mind. And there's no alcohol, there's no smoking, there's no violence through serving the meat and so on. So everything is cleansing. So that creates the environment also that supports people's transformation.
>
> (Ossa)

Cookie also talked about the separation from their everyday life and responsibilities:

> Yeah, I'm rested. You know, I've healed from my chronic fatigue, and I've healed from my years of work, of fatigue. And here I don't have stresses of bill paying, and worrying about everything, and taking care of a house, taking care of animals, and taking care of everyone. I don't even answer my emails. So I'm just... I just detached myself. And this environment allows it, you know.
>
> (Cookie)

The ashram is supported by a large corps of volunteers and is filled with opportunities for selfless service where the demands are not as strenuous as those of the typical work environment. Even for those people who are visiting the ashram on vacation, the lack of the 'profane' in this environment has a significant impact. This is a stark contrast from the values and life-styles of the outside world, especially in the West. As Jamie said, '*When they enter the ashram, there's something that strikes them that is different when they come from the outside*' (Jamie).

To those living at the ashram, the outside world is seen as 'Westernised' and associated with materialism and consumerism. They frame this as a flawed system to live in. Visitors to the ashram escape what they see as 'the negativity' associated with the outside world, often coming to the ashram in physical, mental, emotional and/or spiritual distress, and leaving the ashram feeling healed, transformed or awakened. Time spent at the ashram is different from a typical vacation:

> And he [Swami Vishnudevananda] figured out that when people went on vacation, they usually didn't sleep a lot, they ate unhealthy food, or were partying all night, or whatever it is, and came back from their vacation more depleted than anything else. So what he created is what he called yoga vacations where you'd get up very early during your time here. But you profit so much from this.
>
> (Jamie)

This is especially poignant given the ashram's location (15 minutes by foot) in such close proximity to a large vacation resort, which offers many of the in-dulgences that the ashram does not, including: meat, alcohol, gambling, coffee and other items typical of a luxurious Caribbean vacation. The ashram is seen as free of these things as well as direct pollution, which nurtures a clean, cleansing, authentic environment. Being at the ashram is therefore seen as cleansing one's mind and body and thereby enabling health and wellbeing.

## Gendered experience of blue space

Sivananda is different from the structure of many ashrams in that both men and women can become swamis (as defined above, swamis are Hindu male religious teachers). In fact, Swami Vishnudevananda is recognised as very progressive in that he encouraged women to become swamis. That said, there are still gender differences within the ashram that reflect the wider context of patriarchy and the associated gender ideals/expectations for men and women, as found in both Eastern and Western societies. Yoga was tra-ditionally a male practice where women were forbidden to practise until roughly 150 years ago. Currently, in our contemporary time, at this particu-lar ashram, 70% of the residents and visitors are women. Interviewed partic-ipants suggested that the ashram balances out masculinity and femininity and that yoga philosophy disregards the importance of gender, as people are understood as spiritual beings with human experience (rather than human beings with spiritual experience).

Cookie said that it is easier for women to engage in the practices of the ashram, 'I think it's easier for women to surrender to a space than it is for men', and that '[Men] reason with the thinking of this as not a vacation, [they see being here at the ashram] as punishment' (Cookie). Others suggested that yoga is a more feminine practice in the Western world and, given that men are less in touch with their feminine side, participating in experiences like yoga vacations puts stress on their conceptualisation of gender norms. Lockie pointed out:

> I think you understand that we have to look outside the ashram and see the state of manhood right now in at least the western hemisphere and all over the world, and the state of womanhood, or malehood and femalehood to be more precise, with the question.... I think men have a harder time letting go right now. A harder time letting go. A harder time identifying with the soft side, you know.
>
> (Lockie)

There is also more intrinsic motivation for women to engage in the experience: *'And women want to heal themselves all the time so they can be there for their families'* (Cookie). So, though the ashram is a secluded, sacred place, it still reflects the more prominent patriarchal structure of much of our global society.

## Healing energy

The environment and atmosphere of the ashram facilitates a healing energy that enables the personal transformation and awakening of those who attend. This is evident in the built environment, natural environment (especially the natural blue spaces), social community environment and the spiritual practices that take place within the environment. The participants described how healing happens on the physical, mental, emotional and spiritual levels, and attribute this to the environment of the ashram. As Lockie said: *'They don't come here to eat and get full, they're coming to be fulfilled, to be within their own elements'* (Lockie). Lockie also discussed the different ways the ashram can enable health:

> Yeah. And I've seen addicts come here, and talk about their addiction, and finding their way of teaching yoga and teaching to other addicts. I've seen warriors come here with post-traumatic stress from here to China, and end up being ... teaching yoga to warriors about their experiences. You know, really honest people. Really truthful to their traits. Not denying that they have PTSD, not denying it. Talking about it and bringing it to the open. Letting go of the shame.
>
> (Lockie)

The built environment not only entails the open-air temples and modest built accommodations, but also includes the landscape architecture. Careful attention has been given to the gardens (and flowers within the gardens), the curved paths through the gardens and the colours of both the physical environment and the uniforms worn by staff. These are strategically placed and chosen to reflect simplicity and cultivate the positive energy that resonates throughout the ashram. While the living conditions aren't for everyone, many people find the 'roughing it' environment to be healing. As Lockie explained, '*But the people that stay here, you know, they appreciate the rough environment, the rugged environment, the long nights, the long days, the karma yoga aspect of it*' (Lockie).

The built environment of the ashram plays off of the natural environment it is situated in, as described by Cookie:

> Well, first of all, we have the ocean here. This is huge amounts of energy being poured into this place every day. Whether we like it or not, it's all coming in. And then we're all on coral. And coral's a very special rock. It holds a lot of energy. And we are all on coral here. So coral, it's so much air. So there's so much movement. We're not like grounded. You know, there's a lot of nice energy.
>
> (Cookie)

This merging of the built and natural environment is part of the charm and energy of the ashram. It combines controlled and pristine nature: the gardens, the trees, the beauty of the ocean and the sky, the climate, the beach, the sounds and colours are all healing. Participants discussed the different energies that come from the elements within the environment, such as the ocean, the sky, the trees, the stones and the coral. The ocean in particular is seen as healing for visitors to the ashram. Ossa claimed:

> So that definitely healing of the ocean, like immediately. Long term healing would be the program that we offer – the daily meditation, satsang, the daily yoga classes, the environment, beautiful gardens. The air we breathe, it's clean. There's no pollution here. It's not like other places. And there's a very serene… There is the water dripping in little ponds here and there. There's always like a little oasis where people can sit and have like a quiet moment.
>
> (Ossa)

Cookie summarised the energy of the ashram quite succinctly – '…we'll all say the same thing – the ocean is magical' (Cookie).

It is suggested that elements within the natural environment, such as the ocean or mountains, contribute to and shape the overall energy and atmosphere of the ashram. A participant explained this by contrasting the ashram in the Bahamas to the ashram in Val-Morin, Quebec, Canada. Paula explained the contrast between these two ashrams: '*And here we're lucky because we have the ocean. The ocean also is a very powerful force*', whereas the ashram in Val-Morin is in the Laurentian Mountains, which radiate a '*different force*', as explained by Paula below:

> It's much different. You know, it's in the mountains, in the Laurentians. So you feel a different force. It's a different energy. But you'll notice it. You know, when you go there. Here, it's like very soft and nurturing. You feel that kind of an energy. And then when you go there, you feel more of, I guess, a strong... I don't want to say harsh but you're going to notice it yourself.
>
> (Paula)

Thus, the ocean is understood to cultivate a unique 'soft' and 'nurturing' force that contributes to the mental, physical, emotional and spiritual healing that those staying at the ashram experience.

The community and social environment of the ashram is very accepting and open, helping people to feel included and safe. There is a sense of community and connectedness to something bigger and greater, facilitating a special bond among residents, where they are 'energetically together'. The philosophy of the ashram is to have community members evolve into 'loving' people which, in turn, radiates out from them to others when they leave the ashram to return home. Paula talked about the other residents of the ashram being like brothers and sisters to them, while another said:

> The biggest thing here, even though the beach is beautiful and, you know, there's so many elements here that are very healing, but the main thing is ... sometimes is that connection that they feel to something bigger. Do you know what I mean? Like to... You don't feel isolated. You feel like you're a part of a whole. If that makes sense.
>
> (Paula)

The daily practices and programmes offered at the ashram, including satsang, which are spiritually infused sacred gatherings, made up of chanting, prayer, song and spiritual teachings. Satsangs, yoga practice and the keeping of spiritual diaries, to name a few, are all associated with spirituality which can be healing for those who engage with it. These practices encourage people to live mindfully in the moment, be conscious and present, accept who and what you are, and seek to change and better yourself. Lockie described it as,

> ...not seclusion, or denial, or of ignoring; it's simply being with your own self. This is what the scriptures talk about. It's the ability to be

within your own self and not, you know, looking and being what you want me to be or what they want me to be. Being myself.

(Lockie)

In addition to the spiritual practices of the ashram, people bring with them their own faith, beliefs and associated energy, which add to the healing powers of the ashram. Sivananda Ashram is accepting of all world religions, illustrated in the large signs that greet visitors as they arrive via a three-minute boat ride from Nassau.

Cookie described the energy of the ashram as follows:

And then you've got 50 years of meditation here, every day chanting, every day ... that vibration starts to fill the room. And you can feel it, and you can bathe in it. So yeah, you feel it and you're a part of it. And that's why it's important that everyone come to satsang, and important that everyone come to meditation, and everybody come to everything. Because energetically together, we make it happen.

(Cookie)

The combination of spiritual practices and the length of time they have been practised at this particular location give the ashram a unique 'vibration' where '...*you don't have to take drugs. All you have to do is sit and do karma, and you will be on high*' (Ossa). That said, the energy of the place can be 'too much' for some people who are not ready for it; Jack described one such visitor:

Even though he made the big trip, he couldn't stay because the energy here was so different than the energy ... the underground energy that he was used to in his background. So this guy actually was crying. He was crying, and I held him. Because he just felt the energy of light or the energy was so strong here. It was so far from where he had come from.

(Jack)

**Time extended**

Many people intend to visit the ashram for a short period of time – weeks or months – but then extend their stay or repeatedly come back. A participant illustrated this as, '*Once you taste the honey, you know how delicious and sweet it is*' (Ossa). It was described as a '*healthy addiction*' (Jamie), where people are chasing the high of the energy level they experienced and want to get there again. The amount of time people spend at the ashram depends on their capacity, how much they want to grow, and how much they can handle devoting themselves to the regimen, the structure and contributing to the smooth operation of the ashram. Lockie described how this happens:

Where this place is mainly about karma yoga, about selfless service, you know, and running mostly on volunteers, you know, which, you know, make the place a unique place. A lot of ashrams did not survive on karma yoga. They had to hire a lot of people to run the place, including CEOs and so on. And here, I think the fact that it's volunteering, it grabs people. People enjoy that. And not so much from a pleasure place, from a trust place. There's a certain trust you have in that practice when somebody's volunteering, teaching you. It's not about taking your money.

(Lockie)

Some people are drawn to the ashram and spiritual path as though it were a 'calling'. They define this as seeing one's 'own light' and realizing their purpose through serving others in some capacity and feeling spiritually connected to what they believe in. Paula explained:

I really experienced expansion of understanding things, of knowledge. And I knew that I wanted to come back. So I came back several months later and did the yoga teacher trainer course that they have here. And then that just really opened it up for me. So I eventually... You know, after that, I knew that I wanted to come back as staff. So I came back as staff for one year. And then I just stayed on.

(Paula)

This aligns with the experience of most of the interviewees. They came to the ashram for a visit and then stayed because of divine will or karmic forces pulling them there. *'And that's what karma yogi is. Real karma yoga is not voluntary. You know, it's a path. It's a path of purifying selflessness, purifying the heart'* (Paula).

## Discussion

Part of the value of natural spaces that enable health is that they are separate from everyday life and responsibilities. The ashram is certainly an example of such a natural space for the hundreds of visitors who experience it each week. Most of the participants in the interviews discussed how the natural environment was part of the healing atmosphere of the place. Though it is geographically close to the luxury of a vacation resort (a 15-minute walk), the ashram is sufficiently separate from the 'outside world' to provide the opportunity for physical, mental, emotional and spiritual healing. That said, the ashram can only enable the opportunity for health; it is not inherently healing for everyone.

Participants cited aspects of the built and natural environment as part of the healing power of the ashram. From the gardens, to the temple, to the trees and sky, to the proximity of the ocean – all of these factors play a part in creating an environment that enables health. Participants in this research specifically focused on the ocean as healing, talking about how powerful the energy of the ocean is. One participant specifically talked about how a swim in the ocean worked as a form of therapy; when someone was having a difficult time, this participant would simply tell them to go for a swim and they would come back feeling refreshed and better able to manage their issues. There was disagreement among the participants about the loss of trees in Hurricane Matthew – some feel that the trees helped keep the energy in while others felt that the light and sky exposure increased the energy. Hurricane Matthew had a large impact on the ashram, with the beach platform – where many of the yoga classes are held, being hit hard by the storm and having to be rebuilt as a result.

The social environment is a very important aspect of healing at the ashram. Participants discussed their connections to others within the ashram environment as very close and supportive. This does not mean that there is no conflict between residents and visitors, but that conflict is managed collectively and with peaceful intent. Most conflict results from residents holding onto expectations from the world outside the ashram; letting go of those expectations is noted as an important part of the spiritual healing process of becoming enlightened as a yogi.

The spiritual energy of the ashram was the most commonly discussed factor in the health-enabling powers of this place. Participants talked about the story that the ashram was built on the remains of a temple from the lost city of Atlantis. They claim that this gives the ashram a base of sacred energy that has then been built upon by 50 years of chanting and yoga practice on the site. They believe that these spiritual practices maintain a healing energy at the ashram.

## Conclusions

In this chapter, we have discussed the importance of place in enabling health and specifically, the healing powers of Sivananda yoga ashram in the Bahamas. Looking at this place through the lens of therapeutic landscape theory, we can see that the built/natural environment, social environment and spiritual environment all play a role in creating a health-enabling place. The narratives surrounding spiritual and pilgrimage sites, in particular, have changed over time, with many pilgrims seeking 'healing on a holistic spectrum', not just a physical cure for a disease/ailment (Harris, 2013: 23). At the ashram in particular, the blue spaces of the tropical island environment, together with the spiritual practices engaged in, are especially important.

This study furthers the research on blue spaces and spiritual environments for their healing capabilities. Blue spaces can offer a healing experience for many of the people who interact with them and the spiritual environment amplifies this healing energy. Water has played a role in many spiritual and religious rites for centuries; that the ashram is located at the edge of the ocean is not a mistake. Water holds power and energy that can enable health.

Future research could focus on the health-enabling properties of blue spaces and spiritual environments such as Sivananda Paradise Island from other perspectives, such as that of a guest yogi dependent on a wheelchair or managing some form of chronic illness. Better understanding the health-enabling properties of these environments from the perspective of other contexts would allow for more study of the impact of blue spaces on people with more diverse bodies, as well as how water influences spiritual practices in other traditions. Further, exploring the temporal experience with the ashram would enable a more comprehensive understanding. This may entail a look at how the experience may differ between short-term versus long-term users, and/or single versus repeat users. Finally, there is an obvious need to further explore the spatially adjacent but contrasting worlds of the sacred and profane geographies of the ashram and luxurious casino resort.

## References

Andrews, G.J., Chen, S. and Myers, S. (2014). The 'taking place' of health and well-being: Towards non-representational theory. *Social Science and Medicine,* 108, 210–222.

Bell, S.L., Foley, R., Houghton, F., Maddrell, A. and Williams, A.M. (2018). From therapeutic landscapes to healthy spaces, places and practices: A scoping review. *Science and Medicine,* 196, 123–130.

Bell, S.L., Phoenix, C., Lovell, R. and Wheeler, B.W. (2015). Using GPS and geo-narratives: A methodological approach for understanding and situating everyday green space encounters. *Area,* 47 (1), 88–96.

Collins, D. and Kearns, R. (2007). Ambiguous landscapes: Sun, risk and recreation on New Zealand beaches. In, A. Williams (ed), *Therapeutic Landscapes.* Aldershot, Ashgate, 15–31.

Duff, C. (2012). Exploring the role of 'enabling places' in promoting recovery from mental illness: A qualitative test of the relational model. *Health & Place,* 18, 1388–1395.

Foley, R. and Kistemann, T. (2015). Blue space geographies: Enabling health in place. *Health & Place,* 35, 157–165.

Gesler, W.M. (1992). Therapeutic landscapes: Medical issues in light of the new cultural geography. *Social Science and Medicine,* 34 (7), 735–746.

Gesler, W.M. (1993). Therapeutic landscapes: Theory and a case study of Epidauros, Greece. *Environment and Planning D: Society and Space,* 11 (2), 171–189.

Gladwell, V.F., Brown, D.K., Wood, C., Sandercock, G.R. and Barton, J.L. (2013). The great outdoors: How a green exercise environment can benefit all. *Extreme Physiology and Medicine,* 2 (3), 1–7.

Groenewegen, P.P., van den Berg, A.E., Maas, J., Verheij, R.A. and de Vries, S. (2012). Is a green residential environment better for health? If so, why? *Annals of the Association of American Geographers,* 102 (5), 996–1003.

Harris, A. (2013). Lourdes and holistic spirituality: Contemporary Catholicism, the therapeutic and religious thermalism. *Culture and Religion,* 14 (1), 23–43.

Hartig, T., Mitchell, R., de Vries, S. and Frumkin, H. (2014). Nature and health. *Annual Review of Public Health,* 35, 207–228.

Hoyez, A. (2007). From Rishikesh to Yogaville: The globalization of therapeutic landscapes. In, A. Williams (ed), *Therapeutic Landscapes*. Aldershot, Ashgate, 49–64.

Kearns, R.A. (1993). Place and health: Towards a reformed medical geography. *The Professional Geographer,* 45 (2), 139–147.

Keniger, L.E., Gaston, K.J., Irvine, K.N. and Fuller, R.A. (2013). What are the benefits of interacting with nature? *International Journal of Environmental Research and Public Health,* 10, 913–935.

Lengen, C. (2015). The effects of colours, shapes and boundaries of landscapes on perception, emotion and mentalising processes promoting health and well-being. *Health & Place,* 35, 166–177.

Mitchell, R. (2013). Is physical activity in natural environments better for mental health than physical activity in other environments? *Social Science and Medicine,* 91, 130–134.

Nettleton, S. (2015). Fell runners and walking walls: towards sociology of living landscapes and aesthetic atmospheres as an alternative to a Lakeland picturesque. *The British Journal of Sociology,* 66 (4), 759–778.

Pearce, J., Shortt, N., Rind, E. and Mitchell, R. (2016). Life course, green space and health: Incorporating place into life course epidemiology. *International Journal of Environmental Research and Public Health,* 13, 331–342.

Smith, M. and Kelly, C. (2016). Holistic tourism: Journeys of the self? *Tourism Recreation Research,* 31 (1), 15–24.

Spinney, J. (2006). A place of sense: A kinaesthetic ethnography of cyclists on Mont Ventoux. *Environment and Planning D: Society and Space,* 24, 709–732.

Straughan, E. (2012). Touched by water: The body in scuba diving. *Emotion, Space and Society,* 5, 19–26.

Völker, S. and Kistemann, T. (2011). The impact of blue space on human health and well-being – Salutogenic health effects of inland surface waters: A review. *International Journal of Hygiene and Environmental Health,* 214, 449–460.

Williams, A. (1999). Introduction. In, A. Williams (ed), *Therapeutic Landscapes: The Dynamic between Place and Wellness.* Lanham, MD, University Press of America, 1–11.

Williams, A. (2007). The continuing maturation of the therapeutic landscape concept. In, A. Williams (ed), *Therapeutic Landscapes.* Aldershot, Ashgate, 1–12.

Williams, A. (2017). The therapeutic landscapes concept as a mobilizing tool for liberation. *Medicine Anthropology Theory,* 4 (1), 10–19.

# 9 No ducking, no diving, no running, no pushing

## Hydrophobia and urban blue spaces across the life-course

*Hannah Pitt*

## Introduction

A once notorious sign at UK swimming pools showed a series of cartoons illustrating the rules – no ducking, no pushing, no bombing. I echo this here as emblematic of how water's risky nature influences bodies and behaviour. The series of negatives suggest limitations, converse to freedoms typically associated with blue spaces' therapeutic outcomes. Because constraint deserves attention: unhealthy blue spaces, phobic responses, those who cannot or do not find blue spaces enabling, and how waterscape experiences are modified to reduce risks of disablement (Foley, 2017; Foley and Kistemann, 2015). I focus more on hydrophobia than hydrophilia, meaning not simply fear, but how water repels some from potentially therapeutic places. Where hydrophilia suggests affinity for blue spaces, hydrophobia indicates aversion, pushing away.

Existing research emphasises blue spaces' benefits and those who find them enabling (Foley and Kistemann, 2015; Völker and Kistemann, 2011). But generalisations cannot be based on evidence only from the enabled, as individuals have diverse experiences of the same blue spaces (Finlay et al., 2015). Literature over-looking groups absent from therapeutic environments:

> largely fails to understand why these settings are not emotionally resonant for absent individuals and groups, or how their needs could best be embraced within such atmospheres of care.
>
> (Bell et al., 2018: 11)

We have limited insight to reasons behind differential blue space access. Focusing on beneficiaries skews understanding towards positive perspectives, de-emphasising disabling aspects (Foley and Kistemann, 2015). One recent review found no blue space studies considered water's negative impacts such as drowning (Gascon et al., 2017). Yet even those who find coastal swimming therapeutic acknowledge 'water can disable human movement as well as enable it' (Foley, 2017: 49).

Considering therapeutic landscapes 'darker side', exclusions, contestation and bodies feeling 'out of place' (Bell et al., 2018), my intention is not to rebut more positive emphases but to detail dimensions others note in passing: risks, unpleasant experiences, people not currently using blue spaces for wellbeing. This connects the study of therapeutic spaces with geographies of difference to explore diverse experiences and inequalities (Foley and Kistemann, 2015). Monitoring suggests uneven use of blue spaces, but non-users are under-researched by health geographies. Attending to non-beneficiaries and limits to enablement brings out negative perceptions and experiences, providing evidence for efforts to enable more inclusive access to blue spaces.

Foley and Kistemann highlight that not all waterscapes are blue, but health geographers have neglected murkier waters (2015). The browner palette includes waterways, navigable rivers and canals, constructed and engineered rather than 'natural' watercourses (Pitt, 2018). Although also located in rural areas, my research focused on urban waterways, rarely blue, and for many UK residents, not readily associated with health. Waterways' wellbeing potential is introduced next, followed by findings from research into barriers to access. This is presented as hydrophobia, fear and dislike of blue spaces, which has different emphases across the life-course. The final section questions the potential to overcome hydrophobia to achieve equal access to therapeutic blue spaces.

## Waterways and wellbeing

Evidence of blue spaces' wellbeing value is growing, as is understanding of how they enhance health (Gascon et al., 2017; Völker and Kistemann, 2011). Not all blue spaces are equally enabling, with positive associations between wellbeing and residence near coastal waters less apparent for freshwaters (Wheeler et al., 2012). Understanding such distinctions is stymied by the narrow range of blue spaces researched: rural blue spaces more than urban, coastal more than inland (Völker and Kistemann, 2011, 2015). Some exclude urban waterways for being overly artificial, unnatural environments (Cooper et al., 2017). But urban waterways are used therapeutically, benefits only understood through attention to diverse waterscapes (Pitt, 2018), mirroring a relational perspective on therapeutic green spaces by exploring variations between places and people (Milligan and Bingley, 2007).

Inland waterways are widely accessible as 52% of residents in England and Wales live within 8 km (CRT, 2017b). Houghton and Houghton identify them as alternative therapeutic spaces: 'environments which would usually be "written-off" as negative and harmful' (2015: 281). Often located in edgelands – the indeterminate zone between urban and rural – canals offer many aspects of therapeutic landscapes, what one naturalist calls 'a real change of pace and perspective' (Mabey, 2010: 64). Their therapeutic power comes through countering other urban spaces' disabling properties: 'just to have seen some murky water lapped by non-air-conditioned wind would have set me right' (Mabey, 2010: 17).

The UK's waterway network was created as transport for heavy industry – soon superseded by rail; gradual, seemingly terminal decline followed. By the mid-20th century, surviving waterway landscapes symbolised economic decline: 'smoke stained walls with smashed windows, blocked chimneys sprouting trees between their bricks', neglected by many except those regarding them as 'wet-skips, fit for dumping anything and everything' (Farley and Symmons, 2011: 117–118). The concerted effort of enthusiasts, combined with urban regeneration, restored large parts of the network which is concentrated in former industrial heartlands. Inland waterways are now primarily a recreational network for cycling, walking, water sports and angling. In an average fortnight, 4.3 million adults visit one in England or Wales (CRT, 2017a).

In 2012, most of the network transferred to a new charity, the Canal and River Trust (CRT), marking a transition to management for public benefit through voluntary support. The 2,000 miles managed by CRT comprise canal and navigable river, including significant industrial heritage features (Figure 9.1). The Trust frames waterways as multifaceted contributor to wellbeing (2017b). For example:

> We love and care for your canals and rivers, because everyone deserves a place to escape [...] a world away from every day [...] those struggling with the pressures of day-to-day life have somewhere to rejuvenate themselves.

> (CRT, 2018)

*Figure 9.1* A busy towpath along Regents Canal, London.
Source: Author.

This aligns waterways with escape and restoration associated with therapeutic landscapes (Gesler, 2005). Emerging evidence shows positive effects on subjective wellbeing measures for those who regularly visit a waterway in comparison with the national population (CRT, 2017b: 91). Wellbeing is cited by 72% of recent visitors as motivation, with 56% participating in physical activity and 92% relaxing in a peaceful environment while visiting (CRT, 2017a).

These opportunities to enhance wellbeing suggest waterways as archetypal therapeutic blue spaces, notable for reaching communities far from coastal waters, and being free to access. Over 8 million people in England and Wales, almost 15% of the population, live within 1 km of a CRT-managed waterway, in some areas rising to 70% (CRT, 2017b). Canals' concentration around industry located them among labouring and immigrant communities, urban areas which became heavily disadvantaged following deindustrialisation. The populations in the waterway corridor are typically more ethnically diverse and report poorer health than the national average (CRT, 2017b). Around urban waterways 23% of residents are of Black, Asian and Minority Ethnic (BAME) backgrounds compared to 17% nationally. These groups are less likely to access blue-green spaces or participate in outdoor recreation (Natural England, 2015; Natural Resource Wales, 2015). Their edgelands location makes canals more accessible than 'traditional' therapeutic landscapes for 'more deprived and marginalised populations' (Houghton and Houghton, 2015: 285).

### Under-use and under-representation

Waterways' wellbeing potential is underutilised as significant numbers do not access them. CRT's survey suggests 69% of the national population have not visited one in the last year (2017a). The profile of those accessing waterways is uneven, not reflecting the surrounding population or those with greatest wellbeing needs. Of current visitors (accessed a waterway in the last fortnight) only 45% are female, with an average age of 47 – older than the national population profile (CRT, 2017a). The proportion of BAME visitors is 5% less than the share of the national population, not reflecting the higher proportion of non-white people in urban waterway corridors. Usage reflects UK trends in outdoor recreation and use of blue-green spaces. Long-term monitoring identifies persistent patterns of under-representation of people with disabilities, BAME groups, young people and inner-city residents (Morris and O'Brien, 2011). Blue space use demonstrates inequalities, with 15% of respondents to a UK survey never visiting freshwater blue spaces, with urban dwellers less likely to visit (Bell et al., 2017). A considerable body of literature considers why certain groups are less likely to access outdoor environments (Boyd et al., 2018; Evison et al., 2013; Morris, 2003). But little considers blue spaces, and none focuses on urban waterways so it is unknown whether they present unique barriers. CRT data offer some insight as their national survey of water-way perceptions includes non-users. This found only 43% agree canals are safe places (CRT, 2017a), suggesting safety concerns as a barrier. This illustrates how outdoor spaces, therapeutic to some, are scary to others (Milligan and Bingley, 2007).

## Blue spaces, risk and fear

Perceptions of waterways as unsafe suggest deterrents to using them thera-peutically, or that some consider them disabling. This may be associated with a prevalent public discourse around risk reduction which prioritises water safety (Foley and Kistemann, 2015). Heightened sense of risk could be a sig-nificant barrier, but literature on water activities and outdoor recreation more generally offers a limited perspective on risk. Leisure studies focus on pur-suit of risky activities as affectively positive experiences (Creyer et al., 2003), emphasising extreme adventure enjoyed *because* of high risk (Lipscombe, 2007). Risk is interpreted positively, offering empowerment, sense of escape and rewarding experiences (Burns et al., 2013). But as highlighted in relation to people with disabilities, this excludes some from outdoor recreation. Not everyone deems risk positive, which influences participation (Burns et al., 2013; Lipscombe, 2007).

The relational nature of risk perception was highlighted in Milligan and Bingley's work on woodlands which some find therapeutic, as they identified numerous negative associations inducing fear and anxiety, which deterred some young people (2007). If the same place can be therapeutic or scary depending on person and experience, no landscape can be uncritically re-garded as enabling (ibid.). This has been overlooked in relation to 'natu-ral' environments (Duff, 2011), and blue spaces specifically (Foley, 2017). Many of woodland's risky characteristics – darkness, poor visibility, sense of constraint, fear of attack – feature around urban waterways, as do popu-lar myths of these as scary places. This suggests a need for attention to how risk affects therapeutic use of waterways.

Attention to risk connects with the need to account for differentiated blue space experiences, acknowledging inequalities. Fears which lead some to absent themselves from public spaces have power dynamics which align along lines of age, gender, race and socio-economic advan-tage (Brunton-Smith and Sturgis, 2011). Feeling fearful in public space is linked to vulnerability, one's sense of being able to control life and others' behaviour (Bromley and Stacey, 2012). Spatial features are influential, so over-grown green spaces with signs of neglect feel scarier (Brownlow, 2006). Little research details how water affects risk perception or demographic variations in hydrophobia. But studies of aquatic recreation suggest young males use risk-taking to demonstrate 'dangerous masculinities' while females are more risk averse (Moran, 2011). Gendering may be specific to riskier pursuits, but any water has potential to disable, making risk pertinent to most blue space experiences. As a majority of people perceive waterways as unsafe, they present a useful setting to explore blue spaces risks and how they affect wellbeing.

## Understanding absences from blue spaces

The CRT co-designed research investigating what prevents people accessing inland waterways and how barriers might be addressed (Pitt, 2018). Centred on urban waterways they manage qualitative research focused on four urban locations: Blackburn, Milton Keynes, Tower Hamlets and Leicester. It targeted groups under-represented among users: minority ethnic groups, older adults, young people and parents with young children. Community associations working in the case study locations, such as social housing tenant support teams and youth workers, acted as gatekeepers to approach potential participants. Researchers attended sessions of groups regularly meeting nearby including youth groups, parent toddler sessions and a social group for older men. Families in Leicester and young people in Blackburn were enrolled in sessions run by CRT and local partner organisations to provide facilitated opportunities to experience waterways. People using or familiar with waterways were not excluded, but a majority were not accessing local waterways and 30% had never visited one. Eighty-four people participated, aged 11–85, almost half from minority ethnic groups, just over half female. They joined participatory sessions exploring attitudes and behaviour related to waterways. Researchers attended CRT-led boat trips, volunteering, and walks introducing people to waterways, allowing comparison of attitudes before and after.

Discussions centred on what prevents people accessing local waterways and how they would like to benefit from doing so. Participants of all demographic groups reported positive experiences of urban waterways, describing wellbeing benefits familiar from other chapters: relaxation, physical activity, recreation and active travel. Many properties of therapeutic blue space were associated with waterways, including pleasures of being around water (Foley and Kistemann 2015; Völker and Kistemann, 2011). But even people using them therapeutically noted more disabling properties which I focus on to counter silences in other blue space research. Urban waterways seem to have relatively strong associations with hydrophobia – aversion to or fear of water, causing avoidance, displeasure or disablement in watery environments.

## Dimensions of hydrophobia

The research found hydrophobia constrains behaviour and informs perceptions of waterways as potentially disabling. While not explaining all barriers, it usefully characterises those most significant by prevalence and strength of effect. Participants' waterway hydrophobia demonstrated three core dimensions, the first being displeasure arising from negative environmental qualities. Participants frequently described canals[1] as unpleasant, associated with negative sensory aesthetics such as darkness or smells: '*They need a clean-up, it's quite dirty, the water's not nice, there's rubbish everywhere*' (YM,[2] Blackburn).

Those who had never visited expected mess, sewage, rats or worse:

YF1:  This – low key – is always in the back of my mind, but it's the most concern but it's always in the back of my mind. You know this, on the canal, bodies being found in the canal, stuff like that
YF2:  Yeah that's creepy (Leicester).

They are perceived as unpleasant, attracting unsavoury people:

> On the whole across Blackburn, our canal is an absolute no-no. And that was for all communities not just young people. It's the perception that there's a lot of risk taking behaviour taking place in the area down by the canal.
>
> (Youth worker, Blackburn)

The prevalence of these perceptions demonstrates waterways' characteristics are not wholly positive, repelling some potential users.

A second dimension of waterway hydrophobia is fear, and the perception canals are dangerous:

YF3:  What makes it daunting? Umm… it's just the dark really …
YF1:  It's like dead.
YF3:  It is.
YF4:  No-one really uses that area, so you don't see
HP:  What do you mean?
YF1:  No one really goes to that area so you don't know. If you went to a park…there would be people.
HP:  So there would be nobody there so that would be the
YF4:  No-one to help if something happened.
YF1:  Is there like service?
YF4:  Phone service (Blackburn).

Having never visited a waterway, these women's comments were based on popular perceptions or hearsay. Even those more familiar with canals were influenced by common narratives about bodies in canals or stories of people being pushed in. These may be urban myths or derived from fiction, but are prominent in many people's waterway imagination, signalling them as places of disablement. Across groups and locations, perceived threats to personal safety were a prevalent barrier to accessing waterways. This was strongly associated with images of these as risky environments of unsavoury people or activity: drink or drug use, crime, attackers. Not all risks were specific to blue spaces, for example fear of uncontrolled dogs. But water-ways' characteristics heighten risk or magnify its perception. An attacker can push you into the canal, while a path with few exit routes edged by water inhibits escape.

For a minority fear of water is absolute because they risk drowning:

OM:  Some of the towpaths look a bit narrow and overgrown to me. That and proximity to water puts me off because I don't swim.
HP:  Right. So you're a bit wary of getting too close to the water?
OM:  I'm very wary! (Tower Hamlets).

Those comfortable around water regarded canals as riskier than other blue spaces:

F1:  Someone cycling near the canal immediately you're thinking, you hear all those stories about someone diving in on purpose but getting tangled up in weeds or trollies and not being able to get up, it would put me off but I wouldn't be daft enough to jump in the first place. So it's like...
HP:  But you'd make sure you were careful?
F2:  Yeah, yeah my mum did say that it's not too deep but I know...
F4:  It's something that people associate the canal with though isn't it?
F3:  Yeah of course it is. Well it's water isn't – and you can drown if you fall badly, it's as simple as that isn't it?
F2:  I know someone fell in when they were on a narrow boat and they never came out again because they got trapped under the boat and obviously they had been drinking and that. So I have heard – it's personal stories as well (Blackburn).

Danger was increased by other people:

You know, if a cyclist coming at you, you stand on the edge. If you go further you drop to fall in! So you've got, for instance, the overhead bridge and you're coming round here and a cyclist coming, you know what I mean?

(OM, Tower Hamlets)

Associations with risk-taking activity fuel expectations of encountering 'crazies', 'gangs', 'criminals', 'paedos', 'druggies', 'dealers' and 'alchies', peo-ple regarded as potentially harmful: *'Basically, when I go to the canal and there's drugs or drug dealers that would scare me because I would feel like I'm being attacked'* (YM, Tower Hamlets). The risk is heightened because water compounds other safety concerns: *'The bends are the best place for mug-gers to hide. There's also people being pushed in the canal often'* (OM, Tower Hamlets). The coincidence of water, a topography of confined paths within built-up spaces and perceived or actual presence of nefarious characters heighten hydrophobia around urban waterways, making them risky places even for hydrophiles.

A final dimension of hydrophobia interacts with dislike and fear, as cause and effect of avoidance of waterways. A significant proportion of people not accessing waterways cited lack of awareness of them or how to use them. This man who enjoyed time outdoors had not encountered waterways: *'I'm sure 100%, 80 or 90% of Leicester people they don't know this canal, the same like me, cos I lived here 8 years I don't know as well'* (Leicester).

Others lacked understanding:

> I mean because they don't know these facilities, who own this, whether this wherefore, how they can get access, they don't have information about it so it's like absent from their mind, even if they pass by, they don't hold it, it's just scary, maybe it's just the water, it's the water but they don't realise the leisure and the benefits side of it, just the facilities and what is on offer they don't know.
>
> (M, Leicester)

Several parents familiar with other green-blue spaces had not seen promotion of canals encouraging them to take children. Young participants were particularly likely to be unfamiliar with waterways:

> I think a lot of young people just don't know that they can go on it and it is a public place and you can walk along it and you can go and enjoy it, the canal.
>
> (Youth worker, Blackburn)

This illustrates how low awareness, fear and negative perceptions interact to perpetuate a tendency not to access waterways. People with no habit of visiting canals lack personal experiences so people imagine waterways to be less appealing than they appear through direct experience. Negative perceptions pervade because fear or ignorance prevents hydrophobes from investigating or accessing waterways.

## Hydrophobia across the life-course

Waterway hydrophobia has several interrelated dimensions which repel some from blue spaces; hydrophiles may perceive similar risks but find them less influential than perceived benefits. Whether waterways are found enabling results from interaction between person, place and experience (Conradson, 2005), and the relative strength of health-enhancing and health-limiting aspects (Völker and Kistemann, 2013). Symbolic dimensions of these place experiences suggest sociocultural variations are influential (Bell et al., 2018), something which could only be confirmed through parallel investigations of waterway hydrophobia beyond the UK. The research revealed multiple barriers, varying in intensity from making a canal visit less pleasurable, to absolutely constraining use. Risk lies to the latter end of this spectrum as particularly strong influence on behaviour, while other barriers represent nuisance or inconvenience. Fear was not exclusive to non-swimmers; hydrophiles were wary of waterways because drowning is not the sole risk, and factors beyond swimmers' control threaten their safety. Conversely, inability to swim is not an absolute barrier to enjoying canals. Two Leicester non-swimmers happily joined a boat trip, because the water was calm, the boat was safe and they were accompanied. Blue space risk is not wholly about drowning, rather water's disabling potential in certain contexts, combined with other risks. Attitudes to risk vary (Creyer et al., 2003), so participants judged and responded to waterway risks varyingly. But certain risks were associated with particular groups and fears altered across the life-course, so parents with young children, young people and older people demonstrated hydrophobia with different emphases.

### Parents of young children

Some research identifies lower blue space usage by households with children (Haeffner et al., 2017); reasons are unclear although a focus on beaches identified time, weather, practicalities and finances as barriers (Ashbullby et al., 2013). For parents the prime barrier around waterways was safety risks for infants near unguarded water. Most were relatively active outdoors, often visiting green spaces for infants to play; this group was most amenable to accessing waterways for wellbeing:

> It's just really mindfulness and wellbeing really, exercise for the kids, exercise for us, getting them out for a couple of hours because getting them out is better than being in.
>
> (M, Milton Keynes)

Such places were celebrated for entertaining children while 'doing them good' through fresh air, exercise and nature. But canals were not judged child-friendly because of unprotected edges: 'the risk factor, trying to keep an eye on lots of children in one place' (F, Milton Keynes). Parents were the only group with water safety as the most significant barrier, despite enjoying other waterfronts:

F1: That place at Leighton Buzzard where they've got a splash park and that was brilliant so, something like that, where there is water, they can play in the water, but the water gets drained away so it's fun rather than dangerous.

F2: There's areas where you can go and paddle as well in Haversham so that's quite nice 'cos you know the waters' not that deep, it's a bit – obviously you know, you can't leave them but they're not going to fall, and it's not too deep.

Water is not equally risky with canals problematic as the water edge is ill-defined; several parents suggested barriers would make waterways child-friendly.

Most parents identified wellbeing benefits of children visiting water, but the risk inhibits adults' enjoyment:

> It's nice but you know that you've got to keep – your guard's go to be up, you can't completely relax when you've got the children near the canal. It's a quite narrow path and when you've got three children, trying to walk the three children on that path isn't easy.
>
> (F, Milton Keynes)

Parents liked children to be able to run around in large bounded spaces, so they find narrow towpaths constraining. It was rare for this to prevent ever visiting, but the need for vigilance and restraining children favoured other outdoor spaces. How much risk parents tolerate varies – one mother said, 'I don't feel the danger'. But the prominence of water safety suggests it as consideration for many parents around waterways; hydrophobia centred on unguarded water, the risk of disabling children and how this limits adults' relaxation makes safer blue spaces or dry green spaces more therapeutic.

### Young people

Health and safety is prime concern of older people, and of parents for infants, but as children become teenagers, hydrophobia evolves to include identity-related risks. Young people expressed little affinity for waterways:

YF1: What can you do there?
YF2: Yeah what can you do there?
YF3: Yeah, write that down. There's nothing to do.
YF1: Picnics?
YF3: I mean, you could have a picnic. It's really boring (Blackburn).

This group had the least understanding about waterways, asking if there are footpaths or benches, why people visit. Explaining the recreation available did not help:

> it's just not something a young person would just generally go do, you don't really just go on the side of the canal unless there was some activities on or a festival or something that they had been invited to go along with. I don't think it's something they'll go out of their way to go.
>
> (YF, Blackburn)

This reflects young people's tendency to find natural environments boring, a key barrier to access (Bell et al., 2003; Ethnos, 2005). Teenagers enjoy places offering social interaction and retreat, where they feel ownership (Clark and Uzzell, 2002). For leisure, young participants favoured home, socialising at youth centres, cafes, cinemas, or in town. Those in London were particularly unlikely to favour the outdoors:

> It's always kind of like "oh how long do we have to be outside for. Can we not go back inside? We've got a gym upstairs can't we just use that?" Like they're quite- they want to be indoors, which I don't really know why [...] the majority of people want to be indoors.
>
> (Youth worker)

Elsewhere some young people socialised in parks, or favoured green spaces, but these were minority views, noted as distinguishing them from peers.

Surveys find 12% of under 16s in England never visited a natural environment in the previous year with urban residents less likely to visit (Hunt et al., 2016). One study found 52% of 11–15 year olds would rather do things other than visit the countryside, ranking this as least favoured leisure activity, despite recognising its health benefits (Mulder et al., 2005). Young participants perceived appreciating green-blue spaces and solitary recreation as unusual for their age, associated with a certain type:

> Well I'm not much of a nature person. You know how people feel relaxed when they're around grass and the water and the canal and the water? I'm not that type of person.
>
> (YF, Blackburn)

Others may be deterred from engaging in such behaviour because it is not 'cool':

> I think the peer pressure is a big one. I think the fact that it is such a public place is a big one [...] sometimes I'll upload stuff to our social media accounts and if I upload a canal-themed picture they'll be like "take it down, take it down, our friends will see it, our friends will see it".
>
> (Youth worker, Blackburn)

Peer pressure can encourage young people to sedentary behaviour (Kirby et al., 2013), and deter interest in the countryside (Ethnos, 2005). Outdoor recreation and green space may not fit their lifestyle (King and Church, 2013), so accessing waterways threatens a young person's identity, undermining self- and peer-image as cool or normal. They also risk identification with antisocial behaviour, being 'out of place' in society others view them as a risk (Pain, 2004). Males of BAME heritage in Tower Hamlets avoid canals because of associations with gangs and crime, to evade police attention. Risks specific to young people concerns with preserving identity, personal safety and enjoyable leisure time are not unique to blue spaces, being symptomatic of negative attitudes towards outdoor spaces and recreation. But waterways are so unfamiliar to young people, so associated with risk-taking, that hydrophobia repels them strongly.

### Older people

By post-retirement age the nature of hydrophobia shifts to emphasise personal safety. Older participants were generally well aware of and drawn to waterways, but repelled by concerns for safety and age-related risks. For some poor health and mobility were limiting while the physically able also noted risks:

OM1:  I think people with mobility issues would find some bits of it – I mean there's little slopes even for some people when you go up about a foot and then come down the other side, some of those are quite an impediment for people so they probably wouldn't bother with the canal path.
OM2:  It can be dangerous when there's snow and ice around.
OM1:  And if it's very wet as well that can also be a bit of a pain.
OM3:  But also again with the rather uneven lighting again, that you can get in these places then any irregularity that you can get in ground can become even worse (Tower Hamlets).

A majority of older males in Tower Hamlets enjoyed walking, exploring waterfronts, watching wildlife and learning about heritage – all suiting waterways. But opportunities for enablement were reduced by associations with dangers they perceive in public spaces: '*If you go to a session with the police they will tell you straight off, the most vulnerable are the elderly, they will tell you that*' (OM, Tower Hamlets). They specified encountering groups of young men as a barrier: '*if you need to get past them and you're not quite sure what mood they're in, that would be strong*' (OM, Tower Hamlets). The potential for physical or verbal abuse was off-putting, intimidation not unique to waterways but perceived as heightened in such dark spaces with recesses where attackers could hide. Also remoteness: '*It is very isolating up there. So if something did happen you're a good twenty minutes from any civilisation*' (Tower Hamlets). Older people's caution around blue spaces varies (Finlay et al., 2015); one man resisted his peer's fear of young people and public spaces. Attitudes to risk diverge, but fear of harm is a significant dimension of hydrophobia because it can be an absolute barrier to accessing blue spaces.

### Hydrophobia and risk

Zooming in on different life-course stages suggests waterway hydrophobia as complex, multiple, with varied forms of risk significant to different groups. This should not imply age determines attitudes to blue spaces or risk. But if, as they age a person experiences varied forms of hydrophobia – perhaps around a single waterway – then blue spaces are neither inherently enabling nor disabling. Age is not the only variable influencing behaviour around water: sense of risk increases with vulnerability, a question of power differentials (Bromley and Stacey, 2012; Brownlow, 2006). Infants, older and young people share relative lack of power over public space, hence some common sense of risk in blue spaces. Minority ethnic groups, and females who also feel vulnerable in public spaces (Morris and O'Brien, 2011), expressed safety concerns around waterways, as might people with disabilities who were not included in this research (Burns et al., 2013). Risk perception decreases with familiarity (Lipscombe, 2007), so vulnerability felt by non-users of blue spaces self-perpetuates as people are unlikely to become familiar with a place they fear.

## Diluting the effects of hydrophobia

This exploration of hydrophobia suggests risk and fear as significant barriers to accessing waterways, coalescing around vulnerability, so hydrophobia concentrates among society's least powerful. This suggests any aspiration to attract under-represented groups to blue spaces faces significant challenges embedded in wider inequalities. So can equal access be achieved? Barriers under-represented groups face in relation to outdoor recreation are 'many and complex' (Morris and O'Brien, 2011). Improving facilities is rarely enough as instilling confidence and comfort require affective change (Morris and O'Brien, 2011). This research found some hydrophobia reduced through increased awareness of waterway recreation. For others, a facilitated introduction such as guided walks revealed their image of canals as inaccurate, beginning to establish familiarity without risk. Carefully tailored, led activities can overcome vulnerable groups' fear of unfamiliar outdoor environments, but are resource intensive (Morris and O'Brien, 2011). The effort required to attract hydrophobes to blue spaces prompts another question: *why* pursue equal access? I am reminded of a conversation with a project assistant about our next discussion with young people, how we expected them to be unenthusiastic. She empathised: 'well why *should* they go to a canal?' There is a risk of pursuing normative values around health and outdoor environments, demonising non-users and essentialising blue spaces as good for all, so dominant groups' therapeutic expectations marginalise others (Bell et al., 2018).

Not accessing blue spaces may be a personal choice, but it is one not always freely made, constrained by vulnerability or ignorance of opportunities. Seeking more equal blue space access might meet unmet needs for opportunities to enhance wellbeing as encapsulated by the case of young people. Potentially therapeutic environments are significant for them because of worrying deterioration in mental health (Milligan and Bingley, 2007). Young people (16–24) in the UK increasingly experience anxiety and depression (ONS, 2017); 8–15 year olds' happiness has been at its lowest since 2010 (Children's Society, 2017). Prevalent mental health issues coincide with young people's low levels of engagement with outdoor leisure (King and Church 2013) and physical activity in general (Thompson et al., 2005). Adults' risk aversion combined with fear of abuse constrains children in public spaces (Pain, 2004).

This research did not measure wellbeing, but young participants frequently referred to pressures detrimental to mental health:

> even in our spare time, there's never nothing to do. Even when you don't have homework, you should be revising for your exams because they're really important, do you know what I mean so. As soon as exams start, it's like stress.
>
> (YF, Leicester)

Academic workload and worrying about their future were stressful preoccupations; the first activity one young women imagined doing around waterways was exam revision. These pressures limited teens' leisure time, and ability to relax, threats to their wellbeing suggesting a role for outdoor therapeutic environments. Young people value freedom to roam beyond home and explore outdoor environments (Ethnos, 2005; Natural England, 2010). They recognise how green spaces can benefit wellbeing, and access them for restoration (Milligan and Bingley, 2007). Spaces close to home are particularly valued (Kirby et al., 2013), including some not typically considered natural environments or used by adults (Owens and McKinnon, 2009).

Young people's aspirations for enabling places could align with urban waterways, as confirmed by some young people's descriptions of their value:

> It looks nice. It's, I dunno, it's something that you look at when you just feel like you wanna be alone. You just come out there and see the river or the water. It's really quite a peaceful place to be.
>
> (YM, Tower Hamlets)

A minority use them restoratively:

> The canal helps me to cool off during stressful periods of time and helps me step back and have a better view of life from another perspective. It is calm and helps me forget about all my worries and what is going on no matter how long or brief the time I spent it always makes a vast impact!
>
> (YM, Blackburn)

Young men who walked and cycled along the canal appreciated the chance for solitude, freedom from adults, using the restorative environment to 'clear your head' or 'if you've had a bad day'. They did not idealise canals, perceiving them as dirty and risky, but wellbeing potential could outstrip hydrophobias. These were minority experiences, not reported by young females. Some males visited canals for less enabling pursuits such as drinking, while others remained resolute that waterways offer nothing. But given stigmas around expressing affinity for green spaces or outdoor leisure, and possible discomfort to discuss mental health, appetites for using them restoratively may have remained hidden.

If meeting this latent demand to enhance mental health is a priority for promoting wider use of blue spaces, the challenge for waterway managers is balancing the needs of young people and other hydrophobes. Young people seek places to claim as theirs, but enacting freedom in shared spaces is contested by other users, namely adults threatened by their presence (Bell et al., 2003; King and Church, 2013; Seaman et al., 2010). Waterways will become more risky for adults who find young people intimidating if they become more prominent. Achieving wider access to enabling blue spaces requires solutions which accommodate diverse, conflicting needs. This begins by better understanding those not accessing blue spaces and their hydrophobias.

## Notes

1  Many participants were unfamiliar with the term 'waterways' so discussion referred to canals, the dominant form of waterway in the case study areas.
2  Participants are identified as younger (Y), older (O), female (F) and male (M).

## References

Ashbullby, K.J., Pahl, S., Webley, P. and White, M.P. (2013). The beach as a setting for families' health promotion: A qualitative study with parents and children living in coastal regions in Southwest England. *Health & Place,* 23, 138–147.

Bell, S.L., Foley, R., Houghton, F., Maddrell, A. and Williams, A.M. (2018). From therapeutic landscapes to healthy spaces, places and practices: A scoping review. *Social Science and Medicine,* 196 (Supplement C), 123–130.

Bell, S., Graham, H., Jarvis, S. and White, P. (2017). The importance of nature in mediating social and psychological benefits associated with visits to freshwater blue space. *Landscape and Urban Planning,* 167, 118–127.

Bell, S., Thompson, C.W. and Travlou, P. (2003). Contested views of freedom and control: Children, teenagers and urban fringe woodlands in Central Scotland. *Urban Forestry and Urban Greening,* 2 (2), 87–100.

Boyd, F., White, M., Bell, S. and Burt, J. (2018). Who doesn't visit natural environments for recreation and why: A population representative analysis of spatial, individual and temporal factors among adults in England. *Landscape and Urban Planning,* 175, 102–113.

Bromley, R.D.F. and Stacey, R.J. (2012). Feeling unsafe in urban areas: Exploring older children's geographies of fear. *Environment and Planning A: Economy and Space,* 44 (2), 428–444.

Brownlow, A. (2006). An archaeology of fear and environmental change in Philadelphia. *Geoforum,* 37 (2), 227–245.

Brunton-Smith, I.A.N. and Sturgis, P. (2011). Do neighbourhoods generate fear of crime? *Criminology,* 49 (2), 331–369.

Burns, N., Watson, N. and Paterson, K. (2013). Risky bodies in risky spaces: Disabled people's pursuit of outdoor leisure. *Disability & Society,* 28 (8), 1059–1073.

Children's Society. (2017). The Good Childhood Report, https://www.childrenssociety.org.uk/sites/default/files/the-good-childhood-report-2017_full-report_0.pdf

Clark, C. and Uzzell, D.L. (2002). The affordances of the home, neighbourhood, school and town centre for adolescents. *Journal of Environmental Psychology,* 22 (1), 95–108.

Conradson, D. (2005). Landscape, care and the relational self: Therapeutic encounters in rural England. *Health & Place,* 11 (4), 337–348.

Cooper, B., Crase, L.and Maybery, D. (2017). Incorporating amenity and ecological values of urban water into planning frameworks: evidence from Melbourne, Australia. *Australasian Journal of Environmental Management,* 24 (1), 64–80.

Creyer, E., Ross, W. and Evers, D. (2003). Risky recreation: An exploration of factors influencing the likelihood of participation and the effects of experience. *Leisure Studies,* 22 (3), 239–253.

CRT. (2017a). *Waterway Engagement Monitor 2016–17.* Milton Keynes, Canals and River Trust.

CRT. (2017b). *Waterways and Wellbeing. Building the Evidence Base: First Outcomes Report.* Canal Milton Keynes, Canals and River Trust.

CRT. (2018). https://canalrivertrust.org.uk

Duff, C. (2011). Networks, resources and agencies: On the character and production of enabling places. *Health & Place,* 17 (1), 149–156.

Ethnos. (2005). *What about us? Diversity Review Part 1: Challenging Perceptions: Under-Represented Visitor Needs.* Sheffield, Natural England.

Evison, S., Friel, J., Burt, J. and Preston, S. (2013). Kaleidoscope: Improving support for Black, Asian and Minority Ethnic communities to access services from the natural environment and heritage sectors. *Natural England Commissioned Reports, Number 127.* Sheffield, Natural England.

Farley, P. and Symmons, R. (2011). *Edgelands: Journeys into England's True Wilderness.* London, Jonathan Cape,

Finlay, J., Franke, T., McKay, H. and Sims-Gould, J. (2015). Therapeutic landscapes and wellbeing in later life: Impacts of blue and green spaces for older adults. *Health & Place,* 34, 97–106.

Foley, R. (2017). Swimming as an accretive practice in healthy blue space. *Emotion, Space and Society,* 22 (Supplement C), 43–51.

Foley, R. and Kistemann, T. (2015). Blue space geographies: Enabling health in place. *Health & Place,* 35, 157–165.

Gascon, M., Zijlema, W., Vert, C., White, M.P. and Nieuwenhuijsen, M.J. (2017). Outdoor blue spaces, human health and well-being: A systematic review of quantitative studies. *International Journal of Hygiene and Environmental Health,* 220 (8) 1207–1221.

Gesler, W. (2005). Therapeutic landscapes: An evolving theme. *Health & Place,* 11 (4), 295–297.

Haeffner, M., Jackson-Smith, D., Buchert, M. and Risley, J. (2017). "Blue" space accessibility and interactions: Socio-economic status, race, and urban waterways in Northern Utah. *Landscape and Urban Planning,* 167, 136–146.

Houghton, F. and Houghton, S. (2015). Therapeutic micro-environments in the Edgelands: A thematic analysis of Richard Mabey's the unofficial countryside. *Social Science and Medicine,* 133 (Supplement C), 280–286.

Hunt, A., Stewart, D., Burt, J. and Dillon, J. (2016). *Monitor of Engagement with the Natural Environment: A Pilot to Develop an Indicator of Visits to the Natural Environment by Children.* Sheffield, Natural England.

King, K. and Church, A. (2013). 'We don't enjoy nature like that': Youth identity and lifestyle in the countryside. *Journal of Rural Studies,* 31, 67–76.

Kirby, J., Levin, K.A. and Inchley, J. (2013). Socio-environmental influences on physical activity among young people: A qualitative study. *Health Education Research,* 28 (6), 954–969.

Lipscombe, N. (2007). The risk management paradox for urban recreation and park managers: Providing high risk recreation within a risk management context. *Annals of Leisure Research,* 10 (1), 3–25.

Mabey, R. (2010). *The Unofficial Countryside.* Stanbridge, Little Toller Books.

Milligan, C. and Bingley, A. (2007). Restorative places or scary spaces? The impact of woodland on the mental well-being of young adults. *Health & Place,* 13 (4), 799–811.

Moran, K. (2011). (Young) men behaving badly: Dangerous masculinities and risk of drowning in aquatic leisure activities. *Annals of Leisure Research,* 14 (2–3), 260–272.

Morris, J. and O'Brien, E. (2011). Encouraging healthy outdoor activity amongst under-represented groups: An evaluation of the Active England woodland projects. *Urban Forestry and Urban Greening,* 10 (4), 323–333.

Morris, N. (2003). *Black and Minority Ethnic Groups and Public Open Space.* Edinburgh, Open Spaces.

Mulder, C., Shibli, S. and Hale, J. (2005). Young people's demand for countryside recreation: A function of supply, tastes and preferences? *Managing Leisure* 10 (2), 106–127.

Natural England. (2015). *Monitoring of Engagement with the Natural Environment 2014–2015 Annual Report.* Sheffield, Natural England.

Natural Resource Wales (NRW). (2015). *Wales Outdoor Recreation Survey 2014: Final Report.* Cardiff, Natural Resource Wales.

ONS (2017). Young people's well-being. https://www.ons.gov.uk/peoplepopulation-andcommunity/wellbeing/articles/youngpeopleswellbeingandpersonal finance/2017

Owens, P. and McKinnon, I. (2009). In pursuit of nature: The role of nature in adolescents' lives. *Journal of Developmental Processes,* 4 (1), 43–58.

Pain, R. (2004). Introduction: Children at risk? *Children's Geographies,* 2 (1), 65–67.

Pitt, H. (2018). Muddying the waters: What urban waterways reveal about bluespaces and wellbeing. *Geoforum,* 92, 161–170.

Seaman, P. Jones, R. and Ellaway, A. (2010). It's not just about the park, it's about integration too: Why people choose to use or not use urban greenspaces. *International Journal of Behavioral Nutrition and Physical Activity,* 7 (1), 78.

Thompson, A. Rehman, L. and Humbert, L. (2005). Factors influencing the physically active leisure of children and youth: A qualitative study. *Leisure Sciences,* 27 (5), 421–438.

Völker, S. and Kistemann, T. (2011). The impact of blue space on human health and well-being – Salutogenetic health effects of inland surface waters: A review. *International Journal of Hygiene and Environmental Health,* 214 (6), 449–460.

Völker, S. and Kistemann, T. (2013). "I'm always entirely happy when I'm here!" Urban blue enhancing human health and well-being in Cologne and Düsseldorf, Germany. *Social Science & Medicine,* 78 (0), 113–124.

Völker, S. and Kistemann, T. (2015). Developing the urban blue: Comparative health responses to blue and green urban open spaces in Germany. *Health & Place,* 35, 196–205.

Wheeler, B.W., White, M., Stahl-Timmins, W. and Depledge, M.H. (2012). Does living by the coast improve health and wellbeing. *Health & Place,* 18 (5), 1198–1201.

# Part III

# Blue health inequality and environmental justice

# 10 The shadows of risk and inequality within salutogenic coastal waters

*Sarah L. Bell, Julie Hollenbeck,*
*Rebecca Lovell, Mat White and*
*Michael H. Depledge*

## Introduction

Water has a far-reaching history as a 'sacred substance' (Völker and Kistemann, 2011), as reflected in the plethora of Roman baths, Christian springs and holy wells that continue to hold both spiritual and cultural significance in contemporary society (Gesler, 2003). Perhaps for the first time in the UK, efforts were made to examine the health benefits of inland mineral waters from the 16th century onwards (with the growth in balneology). This scientific interest was extended to the salutogenic potential of coastal waters from the late 18th century with the introduction of coastal sea bathing hospitals intended to treat diseases such as scrofula (Fortescue Fox and Lloyd, 1938). The growth in modern medicine rendered many of the sea-based treatments promoted by these kinds of institutions effectively obsolete in the mid-20th century. However, in recent years interest has emerged in the extent to which coastal environments could help promote public health where non-communicable diseases, including the variety of illnesses associated with a lack of physical activity and poor mental health, are the new priority challenges (Wheeler et al., 2014; White et al., 2017).

Much of the earliest evidence that living in coastal settings encouraged greater levels of recreational physical activity, mainly walking, came from Australia (Ball et al., 2007; Humpel et al., 2004), a country where the majority of the population lives near the coast and weather conditions encourage spending time out of doors. Nevertheless, similar findings have been found in New Zealand (Witten et al., 2008), the US (Gilmer et al., 2003) and the UK (White et al., 2014b). Although there is some evidence that this extra activity may translate into healthier weight, even among children living at the coast (Wood et al., 2016), the evidence is equivocal. In terms of mental health, there is a growing body of research to suggest that living in coastal settings, visiting them frequently or simply having a coastal view from home is associated with increased life satisfaction (Brereton et al., 2008) and decreased risk of anxiety and depression (Nutsford et al., 2016; White et al., 2013a). In part, these mental health benefits may reflect the fact that, on average, visits to coastal settings are associated with greater feelings of mental restoration and feeling relaxed than any other nature or predominantly urban setting (White et al., 2013b). These combined benefits of physical activity and mental health, alongside other factors, help account for findings suggesting that population health tends to be better at the coast (White et al., 2013a, 2014).

Despite this growing evidence of a positive association between human health, wellbeing and coastal living, we know relatively little about the distribution and salience of such benefits – and risks – among different individuals, groups and communities within society. Indeed, we are only just beginning to understand how the physical characteristics of the coast (the particular array of sights, sounds, smells and tactile opportunities) and its shifting sociocultural meanings may play a role in shaping these outcomes for different people. This has important implications for research, practice and policy; if certain people are unable to access 'healthy' coastal encounters – or experience disproportionate levels of risk at the coast – there are fundamental issues of environmental justice and health inequality that need to be addressed.

In this chapter, we adopt a predominantly qualitative, narrative approach to these issues, making space for individuals to reflect on their coastal experiences in their own terms, and exploring the varying embodied, sociocultural and environmental dynamics that shape them. First, we explore the coastal encounters of people living with sight impairment within the UK, based on initial findings from Bell's 'Sensing Nature' study. Second, we draw on Hollenbeck's 'Sea to Me' study of ethno-racial constraints to marine visitation in Miami, Florida, to examine processes of sociocultural segregation and marginalisation at the coast. Finally, stepping back from these detailed case studies, we share insights from work recently undertaken by Depledge, Lovell and colleagues for the UK Government Foresight programme's 'Future of the Sea' initiative, to demonstrate the more widespread challenge of securing and maintaining equitable, healthy encounters with the coast in the face of global climate change. Throughout, we take care to identify both the risks and benefits to human health and wellbeing of varied scales of coastal encounter, and the distribution of such outcomes among different groups and populations across contemporary society.

## Sight loss, sighted norms and the sea: risk and impairment at the coast

Much has been made of the visual properties of aquatic environments (e.g. the colours, light reflections, the sense of space – see White et al., 2014) as a potential mechanism behind some of their benefits, but clearly these are not the only possible sensory encounters. The multisensory immersive properties of the coast have gained increasing research attention over the last few years, with researchers highlighting positive physical and emotional transformations through time spent at the coast, be it within the sea or beside it. Study participants have identified sensations of embodied freedom experienced through physical immersion in the water (Straughan, 2012), a sense of perspective through encountering the 'oceanic horizon' (Bell et al., 2015) and a heightened spatial awareness generated by the 'unceasing mobility' (Ryan, 2012: 9) of the coast, including its rhythmic yet sometimes unpredictable dynamics of light, sound, surface, depth and texture.

Although 'bodies of difference' are mentioned in passing within this work (Foley, 2017; Foley and Kistemann, 2015), there has been relatively little focus on the coastal experiences of people living with sensory impairments, for whom such unceasing mobility and unpredictability may be somewhat less enlivening or freeing. We explore this here in the context of people living with diverse forms of sight impairment, considering how such individuals negotiate the delicate balance between embodied pleasurable immersion and risk exposure that often characterises time spent at the coast. Importantly, we consider how both physical *and* social elements shape the extent to which these individuals experience a sense of risk or wellbeing at the coast.

To do so, we draw on early findings from a two-year in-depth qualitative study, 'Sensing Nature', exploring people's experiences of 'nature' (including but not limited to the coast) with a focus on registered blind and partially sighted adults in England, with congenital or acquired sight conditions, including individuals at diverse life stages. Although the project is still in progress at the time of writing, we reflect here on some initial perspectives emerging within the study interviews – insights which are important in shaping our understanding of risk and joy as simultaneous outcomes (Foley, 2017) of time spent at the coast for varied individuals, groups and communities.

Many of the Sensing Nature study participants described the sense of pleasure experienced through time spent at the coast, noting that although they might not be able to see all – or any – of the details of the seascape, they're *'still picking up on those other things like the sound ... zoning into that sound, maybe just sitting there and just trying to find peace in that moment'* (congenital, partially sighted, male, 40s). Others described the immersive opportunities created by the combined auditory, olfactory and tactile coastal experience; *'Oh the smell of seaweed, I just love the smell of seaweed, you know, ozone and oooooh, just touching the sand, hearing the sea, probably paddling'* (acquired, minimal sight – light perception only, female, 60s).

Such immersive experiences are, however, contingent on particular social, emotional and physical qualities of coastal encounter, together with participants' own embodied skills, dispositions and spatial memory. For example, one participant in her 30s explained that she tends to feel less impaired in nature, particularly along a familiar stretch of coastal path along chalk downland near to her childhood home:

It's probably the environment in which I feel least impaired, compared to say moving around urban environments. So I probably feel most comfortable still walking out on the old trails... I don't need my white stick, you know, at the most I might take a couple of walking poles, and actually it's nice, I probably feel most – in scare quotes – "normal" in nature ... and the great thing about those trails is they're kind of managed enough to be really easy to navigate ... and that's really great because to be honest, I can switch off at a certain level as long as my feet are on that surface.

(Acquired sight impairment, female, 30s)

The combination of familiar paths, clearly defined path textures and boundaries, and the absence of crowds enabled this participant to venture out without her mobility cane – a device that she has embraced to support her mobility through urban environments but which brings with it a sense of being 'different', seemingly singling herself out from fully sighted 'others'.

This is not to say that all participants wanted smooth managed paths to access nature; many emphasised the importance of being able to push themselves, to learn to negotiate more challenging terrains and to have the opportunity to experience a sense of adventure and achievement:

That's the nice thing, because you can sort of feel that [fully sighted] people need to go climbing the Himalayas or doing some massive like Land's End to John O'Groats trip or something, but for me, just walking a mile [along the coastal path] from my house on my own and back, I can get the same thrill.

(Acquired sight impairment, male, 50s)

This participant discussed the time and effort needed to build the confidence, embodied sensory awareness and cane skills to achieve this, and the value in others supporting and being patient through this gradual process of finding a sense of independence in nature. He expressed frustration at the tendency for others to compromise one of the few opportunities he still has to pursue a spirit of exploration or spontaneity:

People's reactions to me make me want to stay away from a lot of things... They can be very overprotective. And very, "Oh no, you don't want to go down that route". Or, "You won't find anything down there". You know, all those sort of negative things. "Look shut up, I'm just exploring!" You know, you can never explore anything.

In this way, the privileging of sighted knowledge over this participant's own sense of autonomy hindered much-valued opportunities to explore and learn about his local coastal environment, while also underestimating the intricate skills he had developed over time to do so.

Several participants touched on the tendency for adverse social encounters to compromise pleasurable or meaningful coastal encounters, largely through the lack of 'sighted' awareness of the dignity of risk and what constitutes appropriate assistance. The example above highlights the detrimental impacts of overprotective, almost infantilising responses. At the other end of the spectrum exists a lack of consideration that people might not be able to see to move out of the way of fast moving objects, be they runners, dogs off leads, cyclists or mobility scooters. This was particularly apparent during a walking interview along a seaside promenade (a shared space for walkers, scooters and cyclists) on the south coast:

> I could go running here on my own if I knew that there wouldn't be any people or that people would just automatically get out of my way or there would be no dog leads... It's that really interesting thing of, often environments aren't disabling for me but the people populating them are.
>
> (Acquired sight impairment, female, 30s)

Others highlighted the tendency for people to stare at 'differential mobilities' (Parent, 2016) developed to negotiate more difficult coastal terrains:

> Walking in the presence of other people, I get quite self-conscious about it, because I do adopt strange strategies for walking. There's the dipping about with the walking pole. And sometimes I will actually bend down and feel what it is I'm going to actually step down into. Or sometimes I'll, I'll get down, you know, sort of, and scramble down something, almost on my backside. Or, you know, scramble up on my hands and, and knees. And I don't want to be watched doing that. And people do, they will stand aside, but they will watch with interest as you do your thing. And I, I really don't like it.
>
> (Congenital sight impairment, female, 60s)

By standing and staring – an act that we are otherwise largely discouraged from doing from a young age – an uneven power dynamic is established between people who are fully sighted and those with sight impairment, rendering the latter almost hyper-visible at a time when full attention is needed to negotiate challenging coastal terrain. Recognising that some 'onlookers' may be unsure about their needs in such situations, participants felt they should simply ask at the outset whether they need assistance, rather than trying to appraise the situation in silence.

These examples demonstrate how opportunities for wellbeing at the coast can be compromised by embedded social (*sighted*) norms, expectations and misinformed notions of risk that fail to recognise, value or cater for alternative ways of sensing or moving through the world (Saerberg, 2010). Such ableist norms and perspectives are often taken for granted and rarely reflected upon by mainstream society (Kitchin, 1998). Yet they can implicitly act to exclude so-called 'bodies of difference' from pleasurable and/or meaningful experiences at the coast as people come to feel out of place through negotiating such settings with adapted mobility strategies. While landscape architects, planners and environmental managers may be aware of the need for more inclusive *physical* environments, the disabling influence of these more intangible social environments – i.e. norms, discourses and sociocultural attitudes – often go unnoticed (Tregaskis, 2004).

## Ethno-racial segregation and the sea: risk and marginalisation at the coast

In light of growing evidence that coastal engagement can contribute to health and wellbeing, it is troubling that these so-called 'healthy blue spaces' (Foley and Kistemann, 2015) are increasingly inaccessible to and/or underutilised by socially, economically or environmentally disadvantaged groups, particularly ethno-racial minorities. Research conducted in the US shows that Black and Hispanic Americans visit aquatic nature settings far less than White or Asian Americans (Leeworthy, 2001). Theories of why race and ethnicity may constrain nature visitation range from structural limitations (e.g. discrimination/racism, proximity, transportation, access points, costs, time), to agentic choices (e.g. preference, disinterest), as well as various sociocultural hypotheses (e.g. marginality, ethnicity, assimilation, discrimination – see Stodolska et al., 2014).

Historically, this research has focused on terrestrial spaces (e.g. local and national parks, mountains, wildlands, 'green spaces' – see Stodolska et al., 2014), with few studies examining use of the recreational marine environment among disadvantaged and minority populations (for example, see Burdsey, 2013). A hallmark study in this area was Wolch and Zhang's (2004) phone survey investigating cultural diversity and recreational marine use and preference in Los Angeles, California; African-Americans perceived fewer constraints to marine access than all other ethno-racial groups in their study, yet visitation still lagged behind other racial groups. They observed that both race and class 'matter', producing class-based, ethno-racial differential recreation patterns. They supposed that cultural preference or 'regional histories of racism and perceptions of places as coded by race/ethnicity in ways that discourage beach use' (Wolch and Zhang, 2004: 438) might account for low rates of visitation by minority populations.

Hollenbeck's (2016) 'Sea to Me' study, conducted in Miami, Florida, explored why the local Black community exhibited low recreational marine environment visitation, when just a generation before beach going was considered mainstream and desirable (Bush, 2016). Participants in this study, predominantly African-American, described the ocean positively (specifically using terms with connotations of health and wellbeing, including freedom, liberation, peace, tranquillity, stress reliever, joyful, therapeutic, reset, comfort, soothing, spiritual) and expressed interest in visiting. With the exception of young men (aged 18–30), participants claimed that neither perceived racism nor discrimination prevented their visitation, noting instead that visiting the sandy beaches and tropical waters just a few miles from their homes simply did not occur to them. Exploring this further, Hollenbeck (2016) concluded that the legacy of historic racism continues to constrain the local Black population from engaging with the marine environment in three historically interconnected ways: intergenerational transmission of the risks associated with coastal and marine visitation, beach-making and segregation, and de-segregation and disconnection.

## Intergenerational transmission

Similar to other coastal cities in the US, historically people of colour were restricted from visiting coastal areas in South Florida until 1965 (and beyond). Visiting some coastal towns without a permit, particularly at night when these towns were closed to Black people ('Sundown towns'), could mean fines, physical force and imprisonment for those who violated these laws and customs. The legacy of these restrictions affected future generations. For instance, one participant in the case study described his lack of coastal visitation as a 'generational curse'. Given the trauma endured by his mother working in the city of Miami Beach in the 1950s–1960s, where she feared being caught after dark, she did not introduce her son to the marine environment. Similar constraints – related to fear and social risk, underpinned by spatio-temporal exclusion – were noted by other study participants, who recalled their parents' reluctance to have them 'out of their sight' as children. This hints at an intergenerational transmission of fear – perhaps an ethno-racial 'strategy' developed to protect their children from abusive encounters with authorities or those outside their neighbourhood.

Between the 1920s and 1970s, Black Miamians were avid beachgoers. However, it was only legal for people of colour to visit one beach, Virginia Key Beach, from 1945 to the mid-1960s (Bush, 2016; Connolly, 2014). Reflecting trends within the broader environmental justice literature (Dean Hardy et al., 2017), this so-called 'Black beach' was undesirable to 'Whites', as it was far from the popular Miami Beach, narrow and steep, and situated off a channel known for treacherous waters and undertows. Demonstrating how systemic sociocultural processes of exclusion and segregation can magnify exposure to physical risks within the marine environment, numerous reports detail Black swimmers drowning off Virginia Key Beach, many of whom were new to ocean engagement. Nearly every participant in the case study knew of someone who had drowned in such outdoor aquatic environments. For this reason, drowning was a primary fear-based constraint to visitation, in addition to fears of dangerous sea life, and the 'unknowability' of the ocean, waves and other ocean phenomena. Negative stories of travel and leisure, transmitted through Black social networks, reinforced a legacy of fear, lack of awareness and inexperience with Miami's coastal regions, and the marine environment in general.

## Beach-making and segregation

Second, the birth of Miami exclusionary segregationist policies and practices, in concert with the purposeful design of its recreational coastal areas as 'America's playground' for White tourists, has resulted in Black residents not living, playing or relaxing 'anywhere near' Greater Miami's coast today (Connolly, 2014: 5, 49). Instead, Black communities are situated in Miami's interior. Many of these communities are criminogenic environments with multiple insecurities and hyper-segregation (>80% racial segregation across numerous indicators). Participants in Hollenbeck's (2016) study were acutely aware of two Miamis, separated by physical bridges: *'when you compare The Beach and you compare it with the inner-city … it's totally two different worlds'* (247). What this participant refers to is mainland Miami, an area they considered the 'real' Miami, versus *'Their*-Ami', which is how some locals referred to the 'other' coastal Miami or The Beach (Connolly, 2014: 118–119). Many participants considered The Beach the domain of tourists and the rich. They described the bridges between Miami's mainland and Miami Beach as delineating the 'real/authentic' versus the 'fake/inauthentic'. In this case, the bridges took a moral or righteous symbolism that constrained access. Reflecting a psychological constraint to visitation, all felt that what lay on the beachside, *Their*-Ami, was material and artificially inflated, while the 'real Miami', where they lived, existed on the landside.

### Desegregation and disconnection

Third, following desegregation in the mid-1960s, local Black recreational marine visitation ebbed. Since all public spaces were legally desegregated, funds to maintain Virginia Key Beach were transferred to a former 'Whites only' beach down the road, Crandon Park (Bush, 2016; Connolly, 2014). Virginia Key Beach soon fell into disrepair and disfavour, ultimately succumbing to closure in the 1980s (reopening in 2008). Though there are dozens of recreational beach areas in South Florida, participants in the case study could only name two: Virginia Key Beach and the new *de facto* Black beach, Haulover. Hollenbeck (2016) concluded that with the closure of Virginia Key Beach, lacking information and direct intervention, Black beachgoers did not alter their sociocultural patterns and practices to include visitation to formerly all-White recreational marine areas that were unfamiliar to them. Therefore, while Black Miamians may have had a rich history associated with Virginia Key Beach and beach going, visitation did not readily transfer to other beach areas. Participants in the case study repeatedly expressed the notion that 'Black folks don't' visit the beach. The acceptance of this notion was generally held across participant groups and demographics, in spite of the fact that a number of participants in the sessions reported visiting marine environments in the past. Lee (2013: 107) described this as passive acceptance 'that African Americans are naturally and inherently disconnected with [the] natural environment' – a pattern seemingly mirrored within regional marine use data demonstrating that local Black visitation is not representative of the Black population within the Greater Miami catchment area (Hollenbeck, 2016).

Given the legacy of exclusionary, isolating and concentrating effects of racial segregation – combined with the intergenerational transmission of attitudes, beliefs and advice passed between family and friends, from generation to generation – today, people of colour in Greater Miami are largely removed from the brief moment in time in which they shared the culturally unifying activity of beach recreation at one of the Nation's few segregation-era beaches. A tapestry of history, politics, capitalism, governance, culture and spatiality shaped a way of being in which community members stopped recreating in or near the ocean to engage in marine activities, with the impetus to do so largely slipping from their individual and collective consciousness. Understanding how best to engage with underrepresented groups, and alleviate their experiences of risk and marginalisation, has become increasingly important as national demographics change and health disparities persist.

## Global challenges for coastal communities: risk and environmental change at the coast

Stepping out from our detailed case studies demonstrating sociocultural dynamics of risk and benefit at the coast, we focus now on the broader challenge of securing and maintaining equitable, healthy encounters with the coast in the face of global environmental change. As noted throughout this chapter, the salutogenic potential of coastal environments is mediated and moderated by myriad factors, from one's individual health and capacities, through to the sociocultural norms of a community or place, and the socio-environmental implications of our changing coastlines. In this penultimate section, we draw on work recently undertaken for the UK Government Foresight programme's 'Future of the Sea' initiative, to illustrate how opportunities for healthy coastal encounters rest on the state of the marine environment itself and the ways in which we respond – nationally and internationally – to the looming challenges of global environmental change. Indeed, as emphasised by Depledge et al.,

> Communities along the coast are on the front line in facing climate change and marine pollution impacts, furthermore their economies are deeply embedded with coastal and other marine activities, making these communities particularly affected. Sea-level rise and extreme weather events, driven by climate change and ecosystem damage, expose coastal communities to flooding events now and in the future, damaging local economies, and threatening health and wellbeing. Continuing pollution of the sea has been underestimated as a threat to the health of coastal dwellers.
>
> (2017: 4)

As such, the beneficial outcomes of environmental change in some areas, including more favourable climates that support increased coastal recreation, will be countered by the heightened risks looming over others. Much environmental change, whether global or local, anthropogenic or natural, poses a direct and specific threat to the health and wellbeing of those who live, work and spend their leisure time in marine and coastal environments. The threats are multiple, operate at a range of geographic scales and can be severe. Many of these threats are compounded and exacerbated by sociopolitical contexts, by processes of marginalisation and by inequities in access to the resources and capacities needed to face and mitigate change. We touch on these here, examining the four main categories of environmental change in the coastal zone that pose significant risks to health and wellbeing globally: climate change, degraded ecosystems, marine and coastal pollution, and development of coastal zones.

First, coastal populations are on the front line in facing *climate change* (Fleming et al., 2015). Some of the most immediate environmental consequences of anthropogenic climate change, such as sea level rise and increased frequency and intensity of extreme weather events, will disproportionally affect coastal regions and communities. The severity of the threat to health is illustrated by recent extreme events such as Hurricane Katrina, which resulted in the deaths of 1800 people and the displacement of over two million people (Hartman and Squires, 2006). Notably, the burden of the disaster fell disproportionately on Black and poorer communities who lacked the social and economic capital to escape (Hartman and Squires, 2006).

Second, *coastal ecosystems* play a vital role in protecting and supporting the health and wellbeing of many millions of people. Inter-tidal habitats and coastal features such as salt marshes, mangroves and reef systems buffer the effects of sea level rises and storms by absorbing wave energy (Möller et al., 2014). The loss of these ecosystems, through climate change, inappropriate development and certain marine industries, threatens the integrity of coastal resilience and the wellbeing of coastal communities. The loss and damage of coastal habitats also threatens wellbeing through reduced economic and recreational opportunity, damage to industries such as tourism, and can result in adverse psychological impacts (Clark et al., 2014).

Third, although water quality is slowly improving in some regions, *marine and coastal pollution* (whether chemical, biological or physical) and the resulting environmental changes pose a direct threat to health and wellbeing. This problem is both significant – for instance there is an estimated 6–12 million tonnes of plastics entering the marine environment *each* year (Jambeck et al., 2015) – and challenging to tackle since the actual sources of pollution can be far upstream to coastal and marine environments. The consequences and impacts of acute and more prolonged pollution events can disrupt industry and livelihoods. For example, in 2011, heavy rain in South Korea resulted in huge quantities of debris, including plastics, washing down the Nakdong River and settling on the beaches of Geoje Island. This event was estimated to have reduced visitor numbers by 63% and to have cost the local economy, to which tourism is a major contributor, approximately US$29–37 million (Jang et al., 2014).

Finally, the *development of coastal zones* to cater for the migration of populations, industries and infrastructure towards the coast can accelerate the processes of environmental change noted above. Ongoing urbanisation and (poorly planned) development of coastal areas damage ecosystems, put pressure on finite resources and leave increasing numbers of people exposed to the impacts of climate change, extreme events and pollution. Key infrastructures such as power stations, ports and sewage treatment plants are often deliberately situated in coastal locations, leaving them vulnerable to sea level rises and flooding events (Neumann et al. 2015). Many coastal communities are relatively geographically isolated, distant from centralised services and without adequate access to vital infrastructure. This peripherality reduces access to education, employment and social opportunities, often leading to processes of exclusion and low self-esteem, poor mental health and harmful behaviours particularly in younger people (Cave, 2010). This again contributes to increased vulnerability of coastal populations to environmental change.

Importantly, these forms of environmental change are interlinked and interdependent. Damage to coastal ecosystems through inappropriate development reduces the capacity of natural defences to mitigate the impacts of other processes of environmental change. Increased sea temperatures due to climate change may exacerbate the impacts of marine pollution (MacLeod et al., 2012). Socio-economic and cultural change must also be considered as both a driver and consequence of environmental change. Recognising that environmental changes are cumulative, often originating at multiple sources and scales (Turner et al., 2015), their influence on the balance of health, wellbeing and risk encountered at the coast can only be understood by examining the wider physical, sociocultural and political contexts in which they are occurring.

## Concluding remarks

In this chapter, we have complemented the growing body of research advocating the salutogenic potential of coastal waters, with detailed case studies demonstrating how the benefits to health and wellbeing for some often emerge within shadows of risk for others. We have reflected on such dynamics at the individual level in the context of embodied risk and impairment, at the community level in terms of the magnification of risk through processes of social segregation and exclusion, and at national and global levels with regard to current and future processes of environmental change and the severe risks they pose for coastal living and livelihoods. The examples presented through the chapter demonstrate how the balance of risk (anticipated and experienced), health and wellbeing can be encountered differently through place and time for diverse individuals and communities, often exacerbating existing patterns of health inequality and environmental (in)justice.

While certain processes of detrimental environmental change may now be irreversible, the social processes that contribute to further change or to unnecessary experiences of risk or health inequality need not be accepted or unchallenged. That experiences of embodied risk continue to be heightened through social processes of exclusion, marginalisation and segregation is a matter of social justice and environmental equity, in which all citizens deserve safe, pollution-free environments as well as equal access to 'health-promoting' coastal environments.

## References

Ball, K., Timperio, A., Salmon, J., Giles-Corti, B., Roberts, R. and Crawford, D. (2007). Personal, social and environmental determinants of educational inequalities in walking: A multilevel study. *Journal of Epidemiology and Community Health*, 61, 108–114.

Bell, S.L., Phoenix, C., Lovell, R. and Wheeler, B.W. (2015). Seeking everyday wellbeing: The coast as a therapeutic landscape. *Social Science and Medicine*, 142, 56–67.

Brereton, F., Clinch, J. P. and Ferreira, S. (2008). Happiness, geography and the environment. *Ecological Economics*, 65, 386–396.

Burdsey, D. (2013). 'The foreignness is still quite visible in this town': Multiculture, marginality and prejudice at the English seaside. *Patterns of Prejudice*, 47 (2), 95–116.

Bush, G.A. (2016). *White Sand Black Beach: Civil Rights, Public Space, and Miami's Virginia Key*. Gainesville: University Press of Florida.

Cave, B. (2010). Health, wellbeing and regeneration in coastal resorts. In, J.K. Walton and Browne, P. (eds.), *Coastal Regeneration Handbook*, Coastal Communities Alliance, 159–174. Available at: http://www.coastalcommunities.co.uk/wp-content/uploads/2015/07/coastal-regeneration-handbook.pdf, last accessed July 1, 2017.

Clark, N.E., Lovell, R., Wheeler, B.W., Higgins, S.L., Depledge, M.H. and Norris, K. (2014). Biodiversity, cultural pathways, and human health: A framework. *Trends in Ecology and Evolution*, 29 (4), 198–204.

Connolly, N.D.B. (2014). *A World More Concrete: Real Estate and the Remaking of Jim Crow South Florida*. Chicago, University of Chicago Press.

Dean Hardy, R., Milligan, R.A. and Heynen, N. (2017). Racial coastal formation: The environmental injustice of colour-blind adaptation planning for sea-level rise. *Geoforum*, 87, 62–72.

Depledge, M., Lovell, R., Wheeler, B.W., Morrissey, K., White, M. and Fleming, L. (2017). *Future of the Sea: Health and Wellbeing of Coastal Communities*. Foresight – Future of the Sea Evidence Review [online]. Available at: https://www.gov.uk/government/uploads/system/uploads/attachment_data/file/639432/Health_and_Wellbeing_Final.pdf, last accessed October 11, 2017.

Fleming, L., Depledge, M., McDonough, N., White, M., Pahl, S., Melanie Austen, Goksoyr, A., Solo-Gabriele, H. and Stegeman, J. (2015). *The Oceans and Human Health*. Oxford Research Encyclopedias: Environmental Science. Oxford, Oxford University Press.

Foley, R. (2017). Swimming as an accretive practice in healthy blue space. *Emotion, Space and Society*, 22, 43–51.

Foley, R. and Kistemann, T. (2015). Blue space geographies: Enabling health in place. *Health & Place*, 35, 157–165.

Fortescue Fox, R. and Lloyd, W.B. (1938). Convalescence on the coast. *The Lancet*, 232, 37–39.

Gesler, W.M. (2003). *Healing Places.* Lanham, Rowman and Littlefield.

Gilmer, M.J., Harrell, J.S., Miles, M.S. and Hepworth, J.T. (2003). Youth characteristics and contextual variables influencing physical activity in young adolescents of parents with premature coronary heart disease. *Journal of Pediatric Nursing,* 18, 159–168.

Hartman, C.W. and Squires, G.D. (2006). *There Is no Such Thing as a Natural Disaster: Race, Class, and Hurricane Katrina.* New York, Routledge.

Hollenbeck, J.A. (2016). Marine Access and Understanding in a Disadvantaged, Urban Coastal Community: Implications for Health, Well-Being and Ecosystem Management. *Ph.D. Thesis,* University of Exeter Medical School.

Humpel, N., Owen, N., Iverson, D., Leslie, E. and Bauman, A. (2004). Perceived environment attributes, residential location and walking for particular purposes. *American Journal of Preventive Medicine,* 26, 119–125.

Jambeck, J.R., Geyer, R., Wilcox, C., Siegler, T.R., Perryman, M., Andrady, A., Narayan, R. and Law, K.L. (2015). Marine pollution. Plastic waste inputs from land into the ocean. *Science,* 347 (6223), 768–771.

Jang, Y.C., Hong, S., Lee, J., Lee, M.J. and Shim, W.J. (2014). Estimation of lost tourism revenue in Geoje Island from the 2011 marine debris pollution event in South Korea. *Marine Pollution Bulletin,* 81 (1), 49–54.

Kitchin, R. (1998) 'Out of Place', 'Knowing One's Place': Space, power and the exclusion of disabled people. *Disability and Society,* 13 (3), 343–356.

Lee, K. (2013). Bourdieuian Analysis on African Americans' Underrepresentation at Parks and Outdoor Recreation. *Dissertation.* Texas A&M University.

Leeworthy, V.R. (2001). *Preliminary Estimates from Versions 1–6: Coastal Recreation Participation. National Survey on Recreation and the Environment (NSRE) 2000.* Washington, DC, US Department of Commerce.

MacLeod, C., Fallon, P.D., Evans, R. and Haygarth, P. (2012). Effects of climate change on the mobilization of diffuse substances from agricultural systems. *Advances in Agronomy,* 115, 41–77.

Möller, I., Kudella, M., Rupprecht, F., Spencer, T., Paul, M., van Wesenbeeck, B.K., Wolters, G., Jensen, K., Bouma, T.J., Miranda-Lange, M. and Schimmels, S. (2014). Wave attenuation over coastal salt marshes under storm surge conditions. *Nature Geoscience,* 7 (10), 727–731

Neumann, B., Vafeidis, A.T., Zimmermann, J. and Nicholls, R.J. (2015). Future coastal population growth and exposure to sea-level rise and coastal flooding – A global assessment. *Plos One,* 10 (3), e0118571.

Nutsford, D., Pearson, A.L., Kingham, S. and Reitsma, F. (2016). Residential exposure to visible blue space (but not green space) associated with lower psychological distress in a capital city. *Health & Place,* 39, 70–78.

Parent, L. (2016). The wheeling interview: Mobile methods and disability. *Mobilities,* 11, 521–532.

Ryan, A. (2012). *Where Land Meets Sea: Coastal Explorations of Landscape, Representation and Spatial Experience.* Farnham, Ashgate.

Saerberg, S. (2010). "Just go straight ahead". *The Senses and Society,* 5 (3), 364–381.

Stodolska, M., Shinew, K., Floyd, M. and Walker, G. (eds) (2014). *Race, Ethnicity, and Leisure: Perspectives on Research, Theory, and Practice.* Leeds, Human Kinetics.

Straughan, E. (2012). Touched by water: The body in scuba diving. *Emotion, Space and Society,* 5, 19–26.

Tregaskis, C. (2004). Applying the social model in practice: Some lessons from countryside recreation. *Disability and Society,* 16 (6), 601–611.

Turner, R., Abernethy, K., Woodhead, A. and Brown, K. (2015). *Health and hidden vulnerability in UK fishing dependent communities.* Environment and Sustainability Institute, University of Exeter.

Völker, S. and Kistemann, T. (2011). The impact of blue space on human health and wellbeing – Salutogenic health effects of inland surface waters: A review. *International Journal of Hygiene and Environmental Health,* 214, 449–460.

Wheeler, B., White, M.P., Fleming, L.E., Taylor, T., Harvey, A. and Depledge, M.H. (2014). Influences of the Oceans on human health and wellbeing. In, R. Bowen, Depledge, M., Carlarne, C. and Fleming, L. (eds), *Oceans and Human Health: Implications for Society and Well-being.* London, Wiley, 4–22.

White, M.P., Alcock, I., Wheeler, B.W. and Depledge, M.H. (2013a). Coastal proximity and health: A fixed effects analysis of longitudinal panel data. *Health & Place,* 23, 97–103.

White, M.P., Pahl, S. Ashbullby, K.J., Herbert, S. and Depledge, M.H. (2013b). Feelings of restoration from recent nature visits. *Journal of Environmental Psychology,* 35, 40–51.

White, M.P., Cracknell, D., Corcoran, A., Jenkinson. G. and Depledge, M.H. (2014a). Do preferences for waterscapes persist in inclement weather conditions and extend to sub-aquatic scenes? *Landscape Research,* 39, 339–358.

White, M.P., Wheeler, B.W., Herbert, S., Alcock, I. and Depledge, M.H. (2014b). Coastal proximity and physical activity. Is the coast an underappreciated public health resource? *Preventive Medicine,* 69, 135–140.

White, M.P., Lovell, R., Wheeler, B., Pahl, S., Völker, S. and Depledge, N.H. (2017). Blue landscapes and public health. In, M. van den Bosch and Bird, W. (eds), *Landscape and Public Health.* Oxford, Oxford University Press.

Witten, K., Hiscock, R., Pearce, J. and Blakely, T. (2008). Neighbourhood access to open spaces and the physical activity of residents: A national study. *Preventive Medicine,* 47, 299–303.

Wolch, J. and Zhang, J. (2004). Beach recreation, cultural diversity and attitudes toward nature. *Journal of Leisure Research,* 36 (3), 414–443.

Wood, S.L., Demougin, P.R., Higgins, S., Husk, K., Wheeler, B.W. and White, M.P. (2016). Exploring the relationship between childhood obesity and proximity to the coast: A rural/urban perspective. *Health & Place,* 40, 126–136.

# 11 Thirst World?

## Linking water and health in the context of development

*Carmen Anthonj and Timo Falkenberg*

### Introduction: water – the foundation of health and development

#### Water as pull factor for human settlement

Worldwide, the pressure on the most precious resource of all – water – is increasing. Ever since the existence of humankind, settlement schemes were developed along waterways, rivers, lakes or any kind of freshwater source, since water was needed to make the settlements strive and prosper. Water is at the core of sustainable development, critical for socio-economic development, energy and food production, healthy ecosystems and human survival itself, serving as the crucial link between society and the environment (United Nations, 2018).

Globally, the world's abundant freshwater resources are not divided equally. Especially in communities in rural arid and semiarid areas of low- and middle-income countries (LMICs), the availability of water is often limited and fragile ecosystems need to be tapped to meet the water needs (Figure 11.1). There, massive population growth along the water courses can be observed with water often being polluted past the point of consumption (Horwitz et al., 2015). Human health depends on the interaction between humans and their surrounding physical, chemical and biological environments. As formulated by Stevens (2010: 268), 'human wellbeing requires a healthy environment, local[ly] and global[ly]'. In the sphere of water and health, research infectious diseases and environmental health risks are prioritised (Hagget, 1994), as well as water scarcity, inadequate water supply, sanitation and hygiene. The interaction between water and health is closely linked to geographical development research, differing in urban and rural areas, with socio-economics disparities, health-related knowledge, risk perceptions and behaviour (Gatrell and Elliott, 2015).

#### Water as the foundation of health

Sufficient water availability is central for human survival. The human hydration requirements are well understood, fluid losses from evaporation, excretion and respiration need to be balanced with fluid intake (Gleick, 1996). White et al. (1972) suggest a minimum per capita water requirement of 3 litres per day in tropical climates, while Howard and Bartram (2003), considering the additional

*Figure 11.1* The meaning and use of water.
Source: Nicol et al. (2015).

hydration requirement of pregnant women, calculated the minimum water volume as 5.5 litres per capita per day under most climatic conditions. The lack of sufficient hydration leads to dehydration, which impairs physical and cognitive functioning, ultimately leading to death. Besides hydration, water is required for the production and preparation of food. Moreover, the availability of water at the household level is fundamental for adequate hygiene behaviour, which also is essential to health. The determinants of hygiene behaviour go far beyond mere availability and accessibility of water, although a positive correlation between frequency of handwashing and available water quantity exists. The underlying factors include sociocultural aspects, economic status, knowledge, education level and risk perception (Gilman et al., 1993).

The health relevance of water reaches beyond its physical use for drinking, food production and hygiene. Over the past decade increasing attention has been placed on the psychological and social meanings of water. Particularly the influence of urban waterbodies on mental health and wellbeing receives increasing attention in academia and urban planning, summarised under the term *Urban Blue* (Völker and Kistemann, 2015). Open blue spaces can act as therapeutic landscapes, where the urban population can relax, recuperate and strengthen their mental and social wellbeing. Water, therefore, has the potential to determine the physical and psychological dimensions of health.

### Health status and development

Intuitively, an association between health status and development is assumed. Generally, the populations of high-income countries (HICs) tend to have better overall health status compared to those of LMICs. Countries with a higher burden of disease show lower levels of economic growth and development than those with a lower disease burden. Nonetheless, the empirical evidence for a causal association between health and development implies a bidirectional relationship (Acemoglu et al., 2003; Eggoh et al., 2015). On the microeconomic level, improvements in health status induce higher productivity through increased physical and mental capacities, including strength, endurance, reasoning and cognitive functioning (Bloom and Canning 2005; Savedoff and Schultz, 2000). On the macroeconomic level, the association between aggregate health indicators and per capita income is well established (World Bank, 1993), indicating a causal correlation between increased income and improvements in health. Increases in per capita income result in a higher purchasing power and thus increasing access to safe food and water, sanitation, education and healthcare (Bloom and Canning, 2005).

It can, however, not be assumed that improvements in health will automatically accelerate development or vice versa. Vollmer et al. (2014) suspect three mechanisms which result in the failure to demonstrate a causal link: unequal distribution of GDP growth, inadequate spending of additional income and failure to translate higher growth to improvements in public service delivery, including health service provision. Eggoh et al. (2015) argue that the accumulation of human capital, which includes health as only one component out of many, is a fundamental determinant of economic growth. Consequently, positive impacts on growth can only be expected when health improvements coincide with improvements in education and social welfare (ibid.). Similarly, when economic growth translates into increased investments in human capital, health improvements can be induced. Thus, institutional structures, income distribution and investment choices strongly mediate the effect (Acemoglu et al., 2003).

## Water as a driver of wellbeing and development

### Applying the sustainable livelihoods framework to capture the value of the blue for health

Water is an important driver in the context of development, and depending on its quality, quantity and use, it serves as livelihood capital that can directly translate into health and wellbeing. The value of water for human health can be distinguished at different levels, ranging from productivity (e.g. for irrigation) to aesthetic (e.g. lakes, river fronts) and recreational (e.g. for swimming), or from the short-term value to fulfil the immediate needs (e.g. hydration, nutrition) to the long-term value (e.g. through sustainable water resource use). One concept that serves to understand the value of water for livelihoods and health at the individual and community levels is the holistic 'Sustainable Livelihoods Framework', an approach rooting in the concept of vulnerability (DFID, 2001). Accordingly, livelihoods include the capabilities, assets and activities required for the means of living in the context of risky living conditions. Assets consist of human capital, natural capital, financial capital, physical capital, and social and cultural capital.

Health, as defined by the World Health Organization (WHO), is a multidimensional concept, and considered to be 'a state of complete physical, mental and social wellbeing and not merely the absence of disease or infirmity' (World Health Organization, 1946: 1). In terms of livelihoods, *good health* is both an asset for and desired outcome of livelihood strategies. Ill-health, on the other hand, makes individuals less productive, constraining their livelihood options (McCartney and Rebelo, 2015; Obrist et al., 2007). Water resources, per se, are an asset in terms of natural capital, upon which particularly individuals of rural societies depend for water supply, cultivation, fishery and livestock watering, thus forming an integral part of their livelihoods (Costanza et al., 1997; Box 11.1). These can be transformed into human capital through the provision of sufficient water for hydration and hygiene, as well as for food production, thus supporting health. Human capital in terms of health-related knowledge, education and risk perception has the potential to reduce the risk of water-related infectious diseases and promote health (Anthonj et al., 2018a; 2019; Rohrmann, 2008; Box 11.4).

Water can contribute to financial capital through the utilisation and sale of the resources provided, contributing to household income and socio-economic status (MEA, 2005). Physical capital with regard to water refers to infrastructures that make water more accessible and safer, including water distribution, irrigation channels and water treatment facilities.

By the individual level, physical capital refers to access to both governmental services (e.g. centralised water supply) and private infrastructure (e.g. household well). Similarly, on the community level, availability of and access to water, sanitation and hygiene (WASH) infrastructure refers to physical capital; however, greater attention is placed on the community coverage of such infrastructure than on individual access. Investments in water infrastructure, both individually and on the community level, improve water quantity and quality, thus translating into human capital through reduced disease risk (ibid.), while also increasing the water use efficiency, potentially generating higher financial capital which in turn contributes to development and health improvement.

The waterscape or *blue space* provides place for recreation, social interaction and salutogenetic health benefits through its physical, aesthetic landscape elements and can thus foster social capital, which entails the social networks and supportive structures of individuals or communities (Foley and Kistemann, 2015; Völker and Kistemann, 2011). Additionally, the availability of water for hygiene contributes to social status, on the individual level, which forms an important determinant of social capital (Bisung and Elliott 2014; Curtis et al., 2009). At the community level, the social fabric of the community as well as potential water conflicts can induce positive or negative health effects.

The cultural value of water influences the relationship between humans and water, affecting usage, management and risk perception (Anthonj et al., 2016; Box 11.4). Cultural capital thus includes habits, behaviours and routines around water, potentially strengthening or threatening human health (Akpabio, 2012). While such habits and behaviours are usually practised on the individual level, cultural capital is fostered on the community (and national) level. Stronger cultural capital through community cohesion positively affects individual cultural capital, and is however mediated by the social fabric (social capital) of the community. The described forms of livelihood assets or capital (natural, financial, physical, social, cultural and human) determine, influence and shape each other, and have influence on and are influenced by access, knowledge and behaviour in the context of water and health (Figure 11.2).

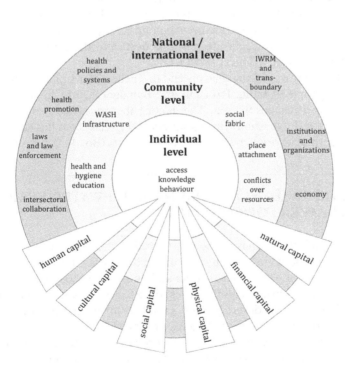

*Figure 11.2* Water, livelihoods and human health at different levels.

At national and international levels, health policies and health promotion, legal frameworks and their implementation, as well as inter- and cross-sectoral collaboration determine health in the context of development, and so do integrated and transboundary water resources' management, institutions and organisations and the global economy.

## Water in a Thirst World: development impediment and health risk

In Sudan, Somalia, Chad, Israel, The Occupied Palestinian Territories, Nigeria, Sri Lanka, Haiti, Colombia and Kazakhstan, shortages of water contribute to poverty. They cause social hardship and impede development. They create tensions in conflict-prone regions. (…) There is still enough water for all of us. But only so long as we keep it clean, use it more wisely, and share it fairly.

Ban Ki-moon, UN Secretary General (2008)

When opening the World Economic Forum in 2008, the former UN Secretary General Ban Ki-Moon summarised water scarcity as one of the most daunting challenges that the world is facing, underlining the conflicts that result from the lack of water, especially in poor nations. He addressed droughts, climate change and the fatal health consequences, the fighting between farmers and herders over water resources. He brought up mismanagement and overuse of water resources, along with population growth, all of which would further accelerate political destabilisation and conflict while increasingly creating a Thirst World:

In the Himalayas, melting glaciers endanger the water supply of hundreds of millions of people in India, Pakistan and Bangladesh. In China, the Yangtze River no longer reaches the sea. The major tributary of the Amazon River in Brazil dried up. Lake Chad with its waters and river system has shrunk to less than one tenth of its original size in the past decades (Box 11.1), putting the water

---

**BOX 11.1 EVIDENCE FROM NIGERIA: WATER SCARCITY AS A DRIVER FOR CONFLICT**

Lake Chad is the main water source in the Northeast of Nigeria and provided livelihoods (fishing, agriculture and livestock farming) for more than 30 million people. The fragile and vulnerable ecosystem had been undergoing a constant change in response to variations in temperature, rainfall, large-scale desertification as well as unsustainable water diversion over the past centuries. As a result, Lake Chad shrunk from 25,000 $km^2$ to only 10% of its original size compared to 1960. At present, the lake is facing acute water shortage and is continuously shrinking, while population growth continues. Shortages of freshwater and increasing water pollution are limiting economic and social development, while at the same time putting an increased pressure on the tributaries.

Moreover, challenges the Lake Chad area is facing include severe political crises and conflict. The lake is shared among four countries with relatively high poverty rates (Cameroon, Chad, Niger, Nigeria), in addition to being located in a sensitive geopolitical area, between Sahara and Sahel, and West and Central Africa. Large-scale social disruptions and regional insecurity arise from increasing water scarcity, declining water quantity and quality along with the shrinkage of the lake.

The use and sharing of the scarce water by different stakeholders is accompanied by increasing competition and conflicts, particularly between farmers and pastoralists fighting over the land and natural resource base for agricultural crop production and livestock grazing (Alemu et al., *forthcoming*).

security of millions of people at risk. While 'water is running out' (Ban Ki-Moon, 2008), now – ten years after the forum – there is more than ever the need to better manage this scarce resource, as the situation has even deteriorated.

### *Water scarcity and human (ill-) health*

Freshwater shortages and water pollution increasingly limit the economic and social development of communities throughout the world (Shiklamanov, 1993). Globally, despite the technological advances of the 21st century, over 1.2 billion people live in regions of physical scarcity, either lacking sufficient water quantity or lacking access to safe water. Five hundred million additional people are approaching the same situation, while another 1.6 billion people face an economic shortage of water (UNDP, 2016). Two-thirds of the global population live under conditions of severe water scarcity at least one month of the year (Mekonnen and Hoekstra, 2016).

A population faces water scarcity, when annual water supplies are below 1,000 m$^3$ per person, while communities with water supplies below 500 m$^3$ exhibit absolute scarcity. Water scarcity is to be differentiated from water stress, which occurs in areas with supplies below 1,700 m$^3$ per person (UN-WATER, 2018).

Economic scarcity occurs when water resources are available but not readily accessible due to lacking infrastructure or high water tariffs. The impacts of water scarcity include severe environmental pollution, declining groundwater levels and conflicts over water use (Hanasaki et al., 2013).

Causes of water scarcity range from natural to anthropogenic. The most important driver of water scarcity is human overuse, above all for agricultural irrigation, accounting for 70% of the global water withdrawal

---

**BOX 11.2 VIRTUAL WATER**

The concept of virtual water was developed by Tony Allan of King's College London to demonstrate the volume of water required for food production and to identify alternative solutions for water-scarce regions. The concept emphasises that trade with products is essentially trade with virtual water, aiming at a more effective use of physical water resources in water-scarce regions (Allan, 2011). It aimed at ensuring food security despite water scarcity in the Middle East and North Africa. The concept of virtual water is closely linked to the Water Footprint, a standardised method to estimate the water volume required for the production of a specific product.

For example, the production process of 1 kg of wheat requires 1,000–2,000 kg of water, while the production of 1 kg of beef requires 16,000 kg of water (Hoekstra, 2003).

In essence, the virtual water concept calls for the rethinking of trade agreements, so that water-scarce regions become net importers of virtual water, rather than exporting large quantities of this essential resource.

---

(United Nations, 2018). The most important natural, environmental cause for water scarcity is drought which is intensified by climate change (ibid., Box 11.2); thus, an expansion of the Thirst World can be expected.

### Access to water in the global development agenda – a human right

To maintain or improve health the mere availability of water resources is not sufficient, but communities and households need to have access to water on their premises, available when needed and supplied free from contamination (WHO/UNICEF, 2017a). One important milestone in terms of water and health has been the recognition by the United Nations General Assembly of the human right to water and sanitation in July 2010 (Resolution 64/292), acknowledging the right of every human being to access to sufficient, safe, acceptable and affordable water for personal and domestic uses. This right encouraged national governments and international organisations to provide financial resources, technology and capacity to ensure affordable access to safe water to all humanity (UN, 2010).

Prior to that, the Millennium Development Goals (MDGs), under MDG 7 (ensuring environmental sustainability), called for halving the proportion of the population without sustainable access to safe drinking water and basic sanitation by the year 2015 (Target 7c). While the target on drinking water was met, the target on basic sanitation was not (in 2015, 2.3 billion people still lacked access to basic sanitation [WHO/UNICEF, 2017]). This fact, along with the need to refine the WASH-related goal to include a target on hygiene, as well as WASH in non-household settings, set the scene of the Sustainable Development Goals (SDGs), defined by the United Nations in 2015.

The now independent water target (SDG 6) aims to achieve universal and equitable access to safe affordable drinking water, adequate and equitable sanitation, and hygiene for all and end open defecation by 2030. The eight sub-targets include the reduction of water pollution, the increase of water use efficiency, the improvement of water management and protection of aquatic ecosystems. Their achievement is designed to contribute to progress across a range of other SDGs, most notably on health (SDG 3), education (SDG 4), economic growth (SDG 8) and the environment (SDGs 14 and 15). Additionally, SDG 6 calls for increased international cooperation as well as more engagement of the local population. The Joint Monitoring Programme (JMP) of the WHO and United Nations International Children's Emergency Fund (UNICEF), established in 2000, monitors and evaluates the development of the WASH infrastructure globally.

Despite the global commitment and relevance of water and health worldwide, currently, 2.1 billion people still lack access to safely managed drinking water services and 4.5 billion people lack safely managed sanitation services (United Nations, 2018).

The JMP differentiates between 'improved' and 'unimproved' water sources and sanitation facilities (Table 11.1); accordingly 'improved' water sources, provided that they are adequately constructed and operated, have the potential to provide safe water (WHO/UNICEF, 2017b). Improved sanitation

*Table 11.1* Classification of drinking water sources and sanitation facilities

|  | *Improved* | *Unimproved* |
|---|---|---|
| Drinking water[a] | Piped water<br>Boreholes or tube wells<br>Protected wells<br>Protected springs<br>Rainwater<br>Delivered water (tanker trucks and small carts)<br>Packaged or delivered water | Unprotected well<br>Unprotected spring<br>Surface water |
| Sanitation[b] | Flush/pour flush to:<br>Piped sewer system<br>Septic tank<br>Pit<br><br>Ventilated improved pit (VIP) latrine<br>Pit latrines with slabs<br>Composting toilets | Pit latrine without slab/ open platform<br>Bucket latrines<br>Hanging toilet/hanging latrine<br>Open defecation |

a  Drinking water from an improved source for which collection time exceeds 30 minutes for a round trip including queuing is not considered improved.
b  Shared or public facilities are not counted as improved.
  Adapted from WHO/UNICEF (2017a).

ensures the hygienic disposal of faeces, avoiding the release of faecal path-ogens into the environment, thus separating excreta from human contact (WHO/UNICEF, 2017b). The technical indicator employed by the JMP assumes that certain 'improved' sources can ensure the delivery of safe water. However, the WHO and UNICEF highlight that improved water sources do not necessarily deliver safe drinking water, but merely reduce the risk of water contamination compared to unimproved sources (WHO/UNICEF, 2000).

The differentiation between improved and unimproved water sources is important for the estimation of health risks; however, these do not define access. The JMP originally defined access according to the distance between the household and the water source, which should not exceed 1,000 m (World Health Organization, 2011). This distance was based on a study by White et al. (1972), who found a direct correlation between the time spent on water collection and household water use. With the updated definition of improved sources (WHO/UNICEF, 2017b), access and distance are now measured by the time the round trip to collect water takes (30 minutes or less). The so-called 'Bradley Curve' (Cairncross, 1987) shows a steep drop in the water volume used, once water collection time exceeds one to three minutes. A second large decrease in household water use is observed when water collection requires more than 30 minutes.

Since the inception of the SDGs, the JMP has further updated the classi-fication, essentially integrating the improved/unimproved classification with service levels, reflecting access. The so-called 'JMP drinking water and sanita-tion ladders' (Table 11.2) define five rungs: safely managed, basic, limited, un-improved and surface water/open defecation (WHO/UNICEF, 2017b). The top three levels of the ladders differentiate between access to improved facilities, while the bottom two rungs distinguish between the utilisation of unimproved facilities (unimproved) and the total lack of facilities (surface water/open defe-cation) (ibid.). Therefore, the JMP assessment moved away from dichotomous classification of improved and unimproved to a more differential classification of service levels. Additionally, the JMP introduced a new third ladder, namely 'JMP hygiene ladder', which defines three levels (basic, limited, no facility). The hygiene service level is defined by the availability of handwashing facilities and soap on premises. Households who have handwashing facilities and soap are defined as 'basic' access, households with handwashing facilities but with-out soap or water fall into the category of 'limited' access, while those without handwashing facilities on premises have no access (no facility).

## Water-related health risks: the Bradley Classification

Where the pressure on and multiple uses of water resources induce the de-terioration of water quality and the reduction of quantity, water-related health risks pose a significant public health threat (Millennium Ecosys-tem Assessment, 2005). The so-called 'Bradley Classification' (White et al. 1972) covers infectious diseases related to water and distinguishes

*Table 11.2* JMP drinking water, sanitation and hygiene service ladders (WHO/
UNICEF, 2017b)

| Drinking water ladder | Sanitation ladder | Hygiene ladder |
| --- | --- | --- |
| Safely managed | Safely managed | |
| Improved source which is located on premises, available when needed and free from faecal and priority contamination. | Improved facility which is not shared with other households and where excreta are safely disposed in situ or transported and treated off-site. | |
| Basic | Basic | Basic |
| Improved source provided collection time is not more than 30 minutes for a round trip including queuing. | Improved facilities which are not shared with other households. | Handwashing facility with soap and water in the household. |
| Limited | Limited | Limited |
| Improved source where collection time exceeds 30 minutes for a round trip, including queuing. | Improved facilities shared between two or more households. | Handwashing facility without soap or water. |
| Unimproved | Unimproved | No facility |
| Unprotected dug well or unprotected spring. | Pit latrines without a slab or platform, hanging latrines and bucket latrines. | No handwashing facility. |
| Surface water | Open defecation | |
| Directly from a river, dam, lake, pond, stream, canal or irrigation channel. | Disposal of human faeces in fields, forest, bushes, open bodies of water, beaches or other open spaces or with solid waste. | |

between four broad, non-exclusive classes of diseases (Table 11.3), providing information on and understanding of transmission, while showing entry points towards disease prevention (Bartram and Hunter, 2015:
21). Waterborne diseases are caused by the ingestion of diverse viral,
bacterial, protozoal and helminthic pathogens in water and are thus determined by the water quality of the water source and at the point-of-
use (Ashbolt, 2004; Hunter, 2010). Water-washed disease transmission is
driven by the availability, access and use of water for personal, food and
domestic hygiene. This category of disease transmission refers to water
as a preventative factor of person-to-person transmission when sufficient
water quantities are available. Water-based diseases cover those pathogens that spend a necessary part of their life cycle in water (Bartram

Table 11.3 Examples of water-related infectious diseases according to the Bradley Classification in the context of development

| Transmission pathway | Example | Disease-causing agent | Symptoms | Link to water | Epidemiological relevance |
|---|---|---|---|---|---|
| Waterborne | Diarrhoea | Bacteria, viruses, parasites; e.g. *E. coli*, Hepatitis A virus, Giardia lamblia. | More than three loose stools per day, secretion of fluids and dissolved salts into gut, dehydration. | Associated with discharged wastewater, inadequate sanitation management and poor drainage, esp. during and after flooding. | Half of all morbidity and a quarter of all mortality in developing countries, esp. in children. |
| Water-washed | Trachoma | Bacteria Chlamydia trachomatis | Visual impairment, permanent corneal damage, irreversible blindness. | Associated with a lack of water for hygiene. Contact with eye discharge of an infected individual, flies, associated with physical and social environment. | Visual impairment in 2.2 million, and blindness in 1.2 million people in poor rural areas. |
| Water-based | Schisto-somiasis (Bilharzia) | Trematode, transmitted by *Schistosoma spp.* | Intestinal and hepatic symptoms, diarrhoea, abdominal pain, enlarged liver, blood in faeces or urine, rash, fever. | Intermediate snail hosts associated with alteration and inappropriate management of water and sanitation systems. | Over 200 million people infected worldwide, 600 million people at risk of infection, and 200,000 deaths annually. |
| Vector-related | Malaria | Protozoa, transmitted by *Plasmodium spp.* or *falcipicarum.* | Fever, jaundice, bleeding disorders, shock, liver failure, encephalopathy death. | *Anopheles* mosquitoes are closely associated with availability of water, flooding, water resources management. | Central public health problem, 225 million cases worldwide and 660,000 deaths annually. |

Adapted from Anthonj et al. (2017), with information from Tulchinsky and Varavikova (2014).

and Hunter, 2015). The fourth of Bradley's categories is the water-related insect vector transmission route, with insects that breed in or biting near water acting as transmission vector.

While the Bradley Classification provides an overview of the different types of water-related diseases, Wagner and Lanoix (1958) highlight the multitude of transmission vehicles of faecal-oral diseases. These diseases are caused by pathogens of faecal origin and largely fall into the category of waterborne or water-washed diseases.

The faecal-oral transmission vehicles are summarised in the 'F-diagram' (Figure 11.3), where faecal pathogens are transmitted via fluids (water), flies (arthropods), fingers (hands), fields (soil) and food to be ultimately ingested by a susceptible host. Fomites are sometimes added to the list of transmission vehicles. In addition to the transmission vehicles, the F-diagram depicts primary and secondary barriers, which prevent transmission if functioning adequately. Primary barriers halt the release of faecal pathogens into the environment and primarily include sanitation and hand hygiene after defecation. Secondary barriers remove faecal pathogens from the environment to prevent ingestion by a susceptible host (Box 11.3). The most important secondary barriers are water treatment (both at source and point-of-use) as well as hygiene behaviour, including hand, food and domestic hygiene. Safe water supply, adequate sanitation and hygiene, referred to as WASH, are crucial preconditions for human health and wellbeing and critical in disease prevention (Black et al., 2003; Clasen et al., 2015; Ejemont-Nwasiaro, 2015; Esrey et al., 1991; Fewtrell et al., 2005; Prüss-Üstün et al., 2014; Waddington et al., 2009; Wolf et al., 2014).

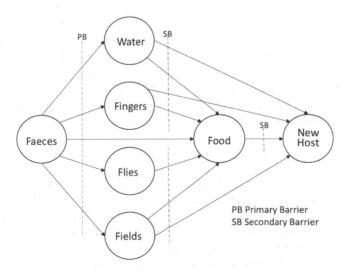

*Figure 11.3* The F-diagram adopted from Wagner and Lanoix (1958).

**BOX 11.3 EVIDENCE FROM INDIA: WASTEWATER-IRRIGATED AGRICULTURE AND HEALTH RISKS**

Ahmedabad is the most populous city of the state of Gujarat and forms the economic engine of the state. Although, along with sustained economic growth, the WASH infrastructure has been expanding in the city, the capacity of the wastewater treatment plants falls short of demand – leading to the discharge of 168 million litres of untreated sewage per day into the Sabarmati River (Ahmedabad Municipal Corporation, 2006). This is particularly alarming, as 90% of the cities' irrigation water requirement is met by surface water (Palrecha et al. 2012).

Unsurprisingly, *Escherichia coli* concentrations of surface water ranged between $9.28 \times 10^5$ and $4.02 \times 10^9$ *E. coli* per 100 ml, far exceeding the standard for unrestricted irrigation (<1,000 *E. coli* per 100 ml) (Falkenberg and Saxena, 2018). Nonetheless, urban farmers in Ahmedabad are reliant on these irrigation water sources, thus exposing themselves, as well as the wider community, to faecal pathogens. The study found a significant increase in the diarrheal disease risk among farmers utilising contaminated irrigation water sources. However, this adverse effect was not restricted to those individuals directly engaged in irrigation and farming but impacts the entire household and the wider community (Falkenberg et al., 2018). Pathogens are transferred by the farmers (hands, clothing, footwear) and their produce into the households and community, as evidenced by an increased probability of in-household drinking water contamination among farmers utilising contaminated irrigation water (ibid.).

This highlights the importance of primary and secondary barriers in the control of water-borne diseases. In the absence of adequate wastewater treatment and disposal, faecal pathogens are (re-)introduced into the community through surface water irrigation, placing the population at risk of water-borne diseases. Even communities that have achieved full sanitation coverage may thus be exposed. It is therefore necessary to include irrigation in the discussion of WASH, essentially transforming WASH into WISH (Water, Irrigation, Sanitation and Hygiene).

Water, sanitation and hygiene cannot be seen in isolation when it comes to human health. Together, they are vital not only for reducing the global burden of disease and improving health, but also for education and economic productivity of populations. Considering the population growth in many LMICs, particularly in urban areas, improvements in WASH in urban blue environments are a necessity for promoting health, wellbeing and healthy development.

Given the high-level commitment manifested in the global development agenda and the human right on water and sanitation, it is no surprise that the implementation of WASH programmes as well as research on the WASH-nexus has been growing extensively over the past decades. Various systematic reviews have demonstrated significant reductions in disease incidence (particularly diarrhoea) through WASH interventions. However, the relative importance of each WASH component is contested. The review of Waddington et al. (2009) demonstrates additional disease reductions when combing sanitation and hygiene interventions with water quality improvements. Esrey et al. (1991) and Fewtrell et al. (2005) recommend combining infrastructure development (hardware intervention) with hygiene education (software intervention), although neither review found evidence for additional impacts from combining interventions. Multiple exposure pathways exist at the WASH-nexus, thus improving one component may reduce exposure but cannot eliminate it (Curtis et al. 2000).

## Preventing disease through health risk perception and healthy behaviour

The interaction with water and the entailed vulnerabilities to acquiring diseases are influenced by health risk perception and health-related knowledge. Risk perceptions have the potential of motivating and shaping health-related behaviour, i.e. the application of protective health measures (Anthonj et al. 2018a). They may reduce or accelerate the risk and exposure to diseases and are therefore valuable, particularly in environments that entail multiple risk factors exposing humans to disease-causing infectious agents (Anthonj et al., 2019).

## The role of health and hygiene education and risk perception

The perceptions of health risks refer to people's intuitive judgements and evaluations of hazards they are or might be exposed to (Rohrmann, 2008), including a multitude of undesirable effects that individuals associate with a specific cause (Rohrmann and Renn, 2000). As described by Bergler et al. (2000), not the objective, but the subjective probabilities relating for example to the individual risk of infection make a given risk a personal risk. Risk perceptions are interpretations of the world, the evaluation of which is influenced by a multitude of individual and societal factors. Such go beyond the classic hazard attributes, represent world views (Rohrmann and Renn, 2000) and are based on experiences, beliefs, attitudes, judgements and feelings, as well as the wider social, cultural and institutional processes (Pidgeon, 1998). Health-related knowledge strongly shapes the perceptions of risk. In many studies, the level of knowledge is measured by the years spent at school. However, experience and knowledge can be acquired elsewhere, e.g. via the

---

**BOX 11.4 EVIDENCE FROM KENYA: RISK PERCEPTIONS REFLECTING ACTUAL RISKS**

An evaluation of risk perceptions among farmers, pastoralists and service sector workers towards infectious disease exposure in a Kenyan wetland revealed the population to possess a high overall level of risk perception regarding the contraction of diseases, despite a low level of formal school education. The population understood exposure to water-related infectious diseases as being driven by physical contact to water during wetland use, characteristics of pathogens and vectors of disease, both in domestic and occupational environments, well.

The risk factors mostly associated with diseases in wetlands included the limited access to improved water supply, sanitation and poor (environmental) hygiene (typhoid fever, diarrhoea, schistosomiasis), agricultural irrigation (malaria), the pastoralists' proximity to livestock (trachoma), the use of agrochemicals (skin and eye diseases), seasonal flooding (malaria, typhoid fever) and droughts (trachoma). Different user groups perceived the use-related risks differently and different (occupational) risks were attributed to different groups. The understanding of disease exposure as due to the intense hydro-social interactions and change present in the fragile semiarid wetland was clear.

The results of this case study show that the risk perceptions reflect the actual risks and shortcomings that such an ecosystem bears. This demonstrates that risk perception studies and resulting recommendations from the grassroots level serve as helpful supportive tools for health-promoting wetland management which requires a sensitive, integrative approach that takes into consideration any and all of the humans, ecology and animals affected (Anthonj et al., 2016, 2018b 2019).

---

radio, newspapers or internet broadcast, health centres, from community health workers or family members. Such sources of information are not easily measurable and quantifiable but are important especially for the illiterate. Although risk perceptions and health beliefs may reflect actual risks (Anthonj et al. 2019, Box 11.4), and although risk perceptions may act as triggers for precautionary action (Wiedemann and Schütz 2005), it should be noted that risk perceptions do not necessarily translate into practice and the engagement in preventive health behaviours (Curtis et al., 2009, 2011). Understanding health and practising healthy behaviour is dependent upon a variety of social, cultural and economic factors, and not limited to infrastructure development and education (Anthonj et al., 2016; Bisung and Elliott 2014; Renner et al., 2008).

## The role of health and hygiene behaviour

Health risk perception are closely linked to and can motivate the practice of health- and hygiene-related behaviour (Curtis et al., 2009; Anthonj et al. 2018a). Behaviour is powerful: *healthy* behaviour, conceptualised for example as secondary barriers to water-related infectious disease transmission (Figure 11.3), may contribute to the prevention of water-related infectious disease transmission. Behaviour is determined by motivation, habits, planning and knowledge, and is largely influenced by the environment and physical factors. The latter include aspects such as the provision, access, availability and cost of water and hygiene, and the proximity and adequacy of sanitation, but also behaviour may underlie seasonal fluctuations. Social factors (including the cultural context of health, health beliefs, norms and traditions) and socio-demographic and socio-economic factors such as wealth, social networks and access to education and information are also highly important in determining health behaviour (Anthonj et al., 2016, 2019; Curtis et al. 2009; Renner et al., 2008). The impact of water on health and development is not merely determined by physical factors of availability, accessibility, quality and quantity, but is strongly mediated by health-related behaviour, which in turn is influenced by a range of social, cultural and economic factors, all of which need to be recognised for health promotion and interventions.

## Outlook

This rechapter presents water as the foundation of human health and a driver of wellbeing and development. Water can act as an impediment to human health and development, as pathogenic and chemical contamination induces health risks, while insufficient water availability constraints personal and domestic hygiene. Such an evaluation goes beyond its physical value (hydration, irrigation, hygiene), as water also holds social and cultural value, affecting mental health, wellbeing and risk perception influencing health-related behaviour. Such salutogenetic effects of waterbodies are increasingly studied under the terms 'therapeutic landscapes' and 'Urban Blue', as these implicit values have been recognised to significantly influence the health status and the quality of life of the population. While research in HICs is increasingly focusing on identifying specific design features that induce positive health effects among different user groups, the concept is scarcely mentioned in the context of LMICs. In order for positive health effects to be induced, waterbodies (and their surrounding environments) need to be clean and aesthetically pleasing, while meeting the requirements of the users (e.g. seating, shading, tranquility, playgrounds, sports facilities). Thus, for LMICs to realise the potential of Urban Blue spaces, first it is necessary to ensure surface waterbodies are clean, which is achieved through the expansion of the WASH infrastructure and the adequate treatment and disposal of wastewater. However, merely focusing on infrastructure development in terms of improved water supply and sanitation is insufficient to reduce the water-borne disease risk. There is an urgent need to couple infrastructure development with social science software (i.e. hygiene and usage behaviour change) interventions to realise the full risk reduction potential. This points to the increased need for health promotion through targeted health programming and education, as well as behaviour change interventions that consider the local and cultural context of water and health.

In light of accelerated population growth, agricultural intensification and climate change, the pressure on global water resources is increasing, driving freshwater scarcity, overexploitation and utilisation of fragile ecosystems. Therefore, honouring the human right to water and sanitation through achieving the United Nations SDG 6 is essential not just to mitigate disease risk and improve health status, but also to enable sustainable development at the economic, social and ecological level. The academic community needs to prioritise research on the nexus of water and health, developing health-based water management systems and designing effective interventions to promote health knowledge, sustainable hygiene behaviour change and the adequate utilisation and management of WASH infrastructure. Such interventions need to be tailored to the specific local context, taking into account the social and cultural value of water, as well as the economic, social and ecological framework conditions. Water can act as driver and impediment for human health and development; it is up to international organisations, national and local governments, as well as civil society to ensure that functioning primary and secondary barriers are in place to mitigate disease risks, while maximising the salutogenetic potential of water.

## References

Acemoglu, D., Johnson, S. and Robinson, J. (2003). Disease and development in historical perspective. *Journal of the European Economic Association*, 1, 397–405.

Ahmedabad Municipal Corporation. (2006). *City Development Plan Ahmedabad 2006–2012*. Ahmedabad, Ahmedabad Municipal Corporation.

Akpabio, E.M. (2012). Water meanings, sanitation practices and hygiene behaviours in the cultural mirror: A perspective from Nigeria. *Journal of Water, Sanitation and Hygiene for Development*, 2 (3), 168–181.

Alemu, W., Anthonj, C., Benner, M.T. and Mohr, O. (*forthcoming*). *A Public Health Analysis in a Protracted Crisis, North East Nigeria*. Abuja, World Health Organization.

Allan, T. (2011). *Virtual Water – Tackling the Threat to Our Planet' Most Precious Resource*. New York, I.B. Tauris.

Anthonj, C., Diekkrueger, B., Borgemeister, C. and Kistemann, T. (2019). Health risk perceptions and local knowledge of water-related disease exposure among Kenyan wetland communities. *International Journal of Hygiene and Environmental Health*, 222 (1), 34–48.

Anthonj, C., Fleming, L., Godfrey, S., Ambelu, A., Bevan, J., Cronk, R. and Bartram, J. (2018a). Health risk perceptions are associated with domestic use of basic water and sanitation services—Evidence from rural Ethiopia. *International Journal of Environmental Research & Public Health,* 15 (10), 2112.

Anthonj, C., Githinji, S. and Kistemann, T. (2018b). The impact of water on health and ill-health in a sub-Saharan African wetland: Exploring both sides of the coin. *Science of the Total Environment,* 624, 1411–1420.

Anthonj, C., Rechenburg, A., Höser, C. and Kistemann, T. (2017). Contracting infectious diseases in Sub-Saharan African Wetlands: A question of use? A review. *International Journal of Hygiene and Environmental Health,* 220 (7), 1110–1123.

Anthonj, C., Rechenburg, A. and Kistemann, T. (2016). Water, sanitation and hygiene in wetlands. The case of Ewaso Narok Swamp, Kenya. *International Journal of Hygiene and Environmental Health,* 219 (7), 606–616.

Ashbolt, N.J. (2004). Microbial contamination of drinking water and disease outcomes in developing regions. Toxicology in the New Century, Opportunities and Challenges. *Proceedings of the 5th Congress of Toxicology in Developing Countries,* 198 (1–3), 229–238.

Ban Ki-moon. (2008). The challenge of securing safe and plentiful water for all is one of the most daunting challenges faced by the world today. United Nations Former Secretary-General's address at the World Economic Forum in Davos, January 24.

Bartram, J. and Hunter, P. (2015). Bradley classification of disease transmission routes for water-related hazards. In, J. Bartram, Baum, R., Coclanis, P.A., Gute, D.M., Kay, D., Mc Fayden, S., Pond, K., Robertson, W. and Rouse, M.J. (eds), *Routledge Handbook of Water and Health.* London and New York, Routledge, 20–37.

Bergler, R., Haase, D. and Schneider, B. (2000). *Irrationalität und Risiko: Gesundheitliche Risikofaktoren und deren naturwissenschaftliche und psychologische Bewertung.* Köln, Kölner Universitäts-Verlag.

Bisung, E., and Elliott, S.J. (2014). Toward a social capital based framework for understanding the water-health nexus. Social Science and Medicine 108 (194–200),

Black, R.E., Morris, S.S. and Bryce, J. (2003). Where and why are 10 million children dying every year? *The Lancet,* 361 (9376), 2226–2234.

Bloom, D.E. and Canning, D. (2005). *Health and Economic Growth: Reconciling the Micro and Macro Evidence.* Harvard, Harvard School of Public Health. Available at: https://pdfs.semanticscholar.org/d773/89f408a288947234a4d154dc9dc4478e90e8. pdf, last accessed June 14, 2017.

Cairncross, S. (1987). Water, women & children: The benefits of water supply. In, J. Pickford and Leedham, B. (eds), *In Developing World Water Vol. II.* London, Grosvenor Press.

Clasen, T.F., Alexander, K.T., Sinclair, D., Boisson, S., Peletz, R., Chang, H.H., Majorin, F. and Cairncross, S. (2015). Interventions to improve water quality for preventing diarrhoea. *Cochrane Database of Systematic Reviews,* 10, CD004794.

Costanza, R., d'Arge, R., de Groot, R., Farber, S.,, Grasso, M., Hannon, B., Limburg, K., Naeem, S., O'Neill, R., Paruelo, J., Raskin, R.G., Sutton, P. and van der Belt, M. (1997). The value of the world's ecosystem services and natural capital. *Nature,* 387, 253.

Curtis, V., Cairncross, S. and Yonli, R. (2000). Domestic hygiene and diarrhoea – Pinpointing the problem. *Tropical Medicine and International Health,* 5 (1), 22–32.

Curtis, V.A., Danquah, L.O. and Aunger, R.V. (2009). Planned, motivated and ha-bitual hygiene behaviour: An eleven country review. *Health Education Research,* 24 (4), 655–673.

Curtis, V., Schmidt, W., Luby, S., Florez, R., Toure, O. and Biran, A. (2011). Hygiene: New hopes, new horizons. *The Lancet Infectious Diseases,* 11 (4), 312–321.

Department for International Development UK (DFID). (2001). *Sustainable Livelihood Guidance Sheet.* London, DFID.

Eggoh, J., Houeninvo, H. and Sossou, G.-A. (2015). Education, health and economic growth in African countries. *Journal of Economic Development,* 40 (1), 93–111.

Ejemot-Nwadiaro, R.I., Ehiri, J.E., Arikpo, D., Meremikwu, M.M. and Critchley, J.A. (2015). Hand washing promotion for preventing diarrhoea. *Cochrane Database of Systematic Reviews,* 2015, 9.

Esrey, S.A., Potash, J.B., Roberts, L. and Shiff, C. (1991). Effects of improved water supply and sanitation on ascariasis, diarrhoea, dracunculiasis, hookworm infection, schistosomiasis, and trachoma. *Bulletin of the World Health Organization,* 69 (5), 609–621.

Falkenberg, T. and Saxena, D. (2018). Impact of wastewater-irrigated urban agriculture on diarrhoea incidence in Ahmedabad, India. *Indian Journal of Community Medicine,* 43 (2), 102–106.

Falkenberg, T., Saxena, D. and T. Kistemann, (2018). Impact of wastewater-irrigation on in-household water contamination. A cohort study among urban farmers in Ahmedabad, India. *Science of the Total Environment,* 639, 988–996.

Fewtrell, L., Kaufmann, R.B., Kay, D., Enanoria, W., Haller, L. and Colford, J.M. Jr. (2005). Water, sanitation, and hygiene interventions to reduce diarrhoea in less developed countries: A systematic review and meta-analysis. *The Lancet Infectious Diseases,* 5 (1), 42–52.

Foley, R. and Kistemann, T. (2015). Blue space geographies: Enabling health in place. *Health & Place,* 35, 157–165.

Gallup, J. and Sachs, J. (2000). The Economic Burden of Malaria, *Working Paper No. 52.* Cambridge, MA, Harvard University Center for International Development.

Gatrell, A.C. and Elliott, S.J. (2015). *Geographies of Health: An Introduction* (3rd ed.). Chichester, Wiley Blackwell.

Gleick, P.H. (1996). Basic water requirements for human activities: Meeting basic needs. *Water International,* 21, 83–92.

Gilman, R.H., Marquis, G.S., Ventura, G., Campos, M., Spira, W. and Diaz, F. (1993). Water cost and availability: Key determinants of family hygiene in a Peruvian shantytown. *American Journal of Public Health,* 83, 1554–1558.

Hagget, P. (1994). Geographical aspects of the emergence of infectious diseases. *Geografiska Annaler,* 76B (2), 91–104.

Hanasaki, N., Fujimori, S., Yamamoto, T., Yoshikawa, S., Masaki, Y., Hijioka, Y.m Kainuma, M., Kanamori, Y., Masui, T., Takahashi, K. and Kanae, S. (2013). A global water scarcity assessment under Shared Socio-economic Pathways – Part 2: Water availability and scarcity. *Hydrology and Earth System Sciences,* 17, 2393–2413.

Hoekstra, A.Y. (ed) (2003). Virtual Water Trade – Proceedings of the International Expert Meeting on Virtual Water Trade. *Value of Water Research Report Series,* No. 12, Delft, IHE Delft.

Horwitz, P., Finlayson, C.M. and Kumar, R. (2015). Human health and the wise use of wetlands – Guidance in an international policy setting. In, C.M. Finlayson, Horwitz, P. and Weinstein, P. (eds), *Wetlands and Human Health*. Dordrecht, Springer, 227–250.

Howard, G. and Bartram, J. (2003). *Domestic Water Quantity, Service Level and Health*. Geneva, World Health Organization.

Hunter, P.R., MacDonald, A.M. and Carter, R.C. (2010). Water Supply and Health. *PLOS Medicine,* 7 (11), 1–9.

McCartney, M.P. and Rebelo, L.-M. (2015). Wetlands, livelihoods and human health. In, C.M. Finlayson, Horwitz, P. and Weinstein, P. (eds), *Wetlands and Human Health*. Dordrecht, Springer, 123–148.

Mekonnen, M.M. and Hoekstra, A.Y. (2016). Four billion people facing severe water scarcity. *Science Advances,* 2 (2), e1500323.

Millennium Ecosystem Assessment. (2005). *Ecosystems and Human Well-Being: Health Synthesis*. Washington, DC, World Resources Institute.

Nicol, A., Langan, S., Victor, M. and Gonsalves, J. (2015). Water-smart agriculture in East Africa. Colombo, Sri Lanka: *International Water Management Institute* (IWMI). CGIAR Research Program on Water, Land and Ecosystems; Kampala, Uganda. Addis Ababa, Global Water Initiative East Africa.

Obrist, B., Iteba, N., Lengeler, C., Makemba, A., Mshana, C., Nathan, R., Alba, S., Dillip, A., Hetzel, M. W., Mayumana, I., Schulze, A. and Mshinda, H. (2007). Access to health care in contexts of livelihood insecurity: A framework for analysis and action. *PLOS Medicine,* 4 (10), e308.

Palrecha, A., Kapoor, D. and Malladi, T. (2012). *Wastewater Irrigation in Gujarat: An Exploratory Study*. International Water Management Institute. Available at: http://www.peopleincentre.org/documents/Wastewater-Irrigation-in-Gujarat. pdf, last accessed, July 8, 2017.

Pidgeon, N. (1998). Risk assessment, risk values and the social science programme: Why we do need risk perception research. *Reliability Engineering & System Safety,* 59 (1), 5–15.

Prüss-Üstün, A., Bartram, J., Clasen, T., Colford, J.M., et al. (2014). Burden of disease from inadequate water, sanitation and hygiene in low- and middle-income settings: A retrospective analysis of data from 145 countries. *Tropical Medicine and International Health,* 19 (8), 894–905.

Renner, B., Schupp, H., Vollmann, M., Hartung, F.-M., Schmälzle, R. and Panzer, M. (2008). Risk perception, risk communication and health behavior change. *Zeitschrift für Gesundheitspsychologie,* 16 (3), 150–153.

Rohrmann, B. (2008). Risk perception, risk attitude, risk communication, risk management: A conceptual appraisal. *The International Emergency Management Society Annual Conference,* 17–19 June 2008, Prague, Czech Republic.

Rohrmann, B. and Renn, O. (2000). Risk perception research. An introduction. In, O. Renn and Rohrmann, B. (eds), *Cross-Cultural Risk Perception. A Survey of Empirical Studies*. New York, Springer, 11–53.

Savedoff, W.D. and Schultz, T.P. (2000). *Wealth from Health*. Washington, DC, Inter-American Development Bank.

Shiklamanov, I.A. (1993). World fresh water resources. In, P.H. Gleick (ed), *Water in Crisis. A Guide to the World's Fresh Water Resources*. New York and Oxford, Oxford University Press, 13–24.

Stevens, P. (2010). Embedment in the environment: A new paradigm for well-being? *Perspectives in Public Health*, 130 (6), 265–269.

Tulchinsky, T.H. and Varavikova, E. (eds) (2014). *The New Public Health*. Amsterdam, Academic Press, 884.

United Nations. (2010). *Resolution 64/292*, United Nations. Available at: http://www.un.org/es/comun/docs/?symbol=A/RES/64/292&lang=E, last accessed June 6, 2017.

United Nations. (2018). *Water*. Available at: http://www.un.org/en/sections/issues-depth/water/, last accessed June 6, 2017.

United Nations Development Programme. (2016). *2016 Human Development Report. Human Development for Everyone*. New York, United Nations Development Programme.

United Nations Water. (2018). Water scarcity. Available at: http://www.unwater.org/water-facts/scarcity/, last accessed June 6, 2017.

Völker, S. and Kistemann, T. (2011). The impact of blue space on human health and well-being – Salutogenetic health effects of inland surface waters: A review. *International Journal of Hygiene and Environmental Health*, 214 (6), 449–460.

Völker, S. and Kistemann, T. (2015). Developing the urban blue: Comparative health responses to blue and green urban open spaces in Germany. *Health and Place*, 35, 296–205.

Vollmer, S., Harttgen, K., Subramanyam, M.A., Finlay, J., Klasen, S. and Subramanian, S.V. (2014). Association between economic growth and early childhood undernutrition: Evidence form 121 Demographic and Health Surveys from 36 low-income and middle-income countries. *Lancet Global Health*, 2 (4), e225–e234.

Waddington, H., Snilstbeit, B., White, H. and Fewtrell, L. (2009). Water, sanitation, and hygiene interventions to combat childhood diarrhoea in developing countries. *The International Initiative for Impact Evaluation* (3ie), 3ie Synthetic Reviews. Available at: http://www.3ieimpact.org/media/filer_public/2012/05/07/17-2.pdf, last accessed June 6, 2017.

Wagner, E.G. and Lanoix, J.N. (1958). Excreta disposal for rural areas and small communities. *World Health Organization Monograph Series No.39*. Geneva, World Health Organization.

White, G.F., Bradley, D.J. and White, A.U. (1972). *Drawers of Water: Domestic Water Use in East Africa*. Chicago, University of Chicago Press.

Wiedemann, P.M. and Schütz, H. (2005). The precautionary principle and risk perception: Experimental studies in the EMF area. *Environmental Health Perspectives*, 113 (4), 402–405.

Wolf, J., Prüss-Ustün, A., Cumming, O., Bartram, J., Bonjour, S., Cairncross, S., Clasen, T., Colford, J.M. Jr., Curtis, V., De France, J., Fewtrell, L., Freeman, M.C., Gordon, B., Hunter, P.R., Jeandron, A., Johnston, R.B., Mäusezahl, D, Mathers, C., Neira, M. and Higgins, J.P. (2014). Assessing the impact of drinking water and sanitation on diarrhoeal disease in low- and middle-income settings: Systematic review and meta-regression. *Tropical Medicine and International Health*, 19 (8), 928–942.

World Bank. (1993). *World Development Report 1993: Investing in Health*. Washington, DC, World Bank.

World Health Organization. (1946). Preamble to the Constitution of the World Health Organization as adopted by the International Health Conference, New York, 19–22 June, 1946.

World Health Organization. (2011). *Guidelines for Drinking-Water Quality* (4th ed.). Geneva, World Health Organization.

WHO/UNICEF. (2000). *Global Water Supply and Sanitation Assessment 2000 Report*. New York, United Nations Children's Fund & World Health Organization.

WHO/UNICEF. (2017a). *JMP Methodology – 2017 Update & SG Baseline*. Geneva, United Nations Children's Fund & World Health Organization.

WHO/UNICEF. (2017b). *Progress on Drinking Water, Sanitation and Hygiene – 2017 Update and SDG Baselines*. Geneva, United Nations Children's Fund & World Health Organization.

# 12  Wellbeing and the wild, blue 21st-century citizen

*Sarah Atkinson*

## Introduction

The last ten years have seen a marked rise in the numbers of people choosing to swim in lakes, rivers or the sea, an activity variously labelled as open water swimming, sea swimming or wild swimming. This chapter considers reasons for this trend in the context of the United Kingdom through two deceptively simple modes of explanation: the individual benefits and attractions of wild swimming and wider contemporary values and fashions encouraging the growth of this activity. The dominance in contemporary health promotion of calls for greater exercise as a route to fitness, improved health and longevity appears to offer an easy explanation for why wild swimming has seen a surge in popularity: the health and wellbeing gains that are possible from swimming appear self-evident and, in the context of exhortations to exercise, provide sufficient explanation for the increase in uptake. The focus, therefore, of the chapter is directed less to why some people's preferred response to such exhortations to exercise is to take up swimming, but to why their response is to choose a variant of swimming that is outdoors and in open and relatively 'wild' bodies of water. After first describing the phenomenon of the growth in wild swimming, the chapter reviews arguments for the benefits that wild swimming may bring for health and wellbeing. This prompts a subsequent section arguing that the literature on wild swimming, both academic and autobiographical, shares a number of elements that effectively reveal convergence in thinking about wild swimming. The chapter then explores how this dominant approach to wild swimming may be understood as embedded in wider historical and contemporary narratives which serve to include only some aspects of wild swimming while excluding others. The final sections bring the benefits and the wider narratives together to briefly reflect on this dominant orientation in understanding modern enthusiasms for wild swimming.

## The wild swimming phenomenon

Wild swimming in the United Kingdom has been a phenomenon of the 2010s. Earlier times have seen outdoor swimming come in and out of fashion including a mediaeval interest in bathing, the rise in the 18th and 19th centuries of seaside resorts (Parr, 2011) and the expansion in the 1920–1930s of lidos and swimming pools across Europe as concerns moved from cleansing to health and the attention to outdoor sport associated with fascism (Bolz, 2012; Dogliani, 2000; Marino, 2010; Pussard, 2007). What is, perhaps, different with the contemporary vogue is the explicit engagement in rather unstructured spaces, the wild places of sea, rivers and lakes. The swimming activity undertaken can vary hugely, from the demanding exertions of competitive or long-distance swimming through to the gentle routine and short swims or dips. In the United Kingdom, the growth in participation came to the attention of the media in the mid-2010s, leading to a flurry of newspaper coverage in 2015 (see, for example, Etherington, 2015; Laville, 2015) and a concomitant decrease in numbers swimming in public pools (see Rhodes, 2016). Overviews of the phenomenon consistently date its appearance to around 2006–2008 and as associated with the first organised one mile swim in Lake Windermere in 2006 and the inclusion of the triathlon and the swimming marathon as Olympic events in the 2008 Beijing games (Laville, 2015). The first organised swim in Lake Windermere in 2006 prompted the launch of the Outdoor Swimming Society with just 300 members (*https://www.outdoorswimmingsociety.com/*); by 2015, the membership numbered 23,000. The first formal mass swimming event of the society was the Great Swim in Lake Windermere in 2008, with 3,000 participants. This has become an annual event and not only at Windermere but also in other locations; by 2015, over 20,000 registered participants engaged in five annual Great Swim events. In addition to these large, high-profile events, in 2015 there were over 170 smaller wild swimming events of varying distances and locations being run in the United Kingdom. There are also a few longer swimming events on offer, including a 10-km marathon swim in Lake Windermere for which places for the 2016 event were almost sold out within a few days of registration opening (Laville, 2015).

The growth in participation and interest has supported a dedicated magazine launched in 2011 as '*H2Open*' and later rebranded in 2017 as '*Outdoor Swimmer*' which is now published monthly (*https://outdoorswimmer. com/*). The Outdoor Swimming Society has mapped locations across the country suitable for wild swimming, which are available on their website or through the publication *Wild Swim* (Rew, 2008) and there are several online sites where swimmers provide detail and advice about specific sites (for example, *www.wildswimming.co.uk*). The rise in wild swimming participation has been accompanied by a recent outpouring of autobiographical writing about the experience (Fitzmaurice, 2017; Heminsley, 2017; Lee, 2017; Minihane, 2017; Peters, 2014; Shadrick, 2017; Wardley, 2017); this emerging trend bridges nature writing and autobiography and has already been termed 'waterbiography' by Jenny Landreth in relation to her own account, *Swell* (Landreth, 2017). This trend of waterbiography was initiated at the turn of the millennium with Roger Deakin's autobiographical account *Waterlog* (Deakin, 1999), an account which is now regarded as a classic of the genre of new-nature writing (Moran, 2014).

### The benefits of wild swimming

Across a number of disciplines and range of methods, a body of evidence supports a narrative that being outdoors and particularly in the green and blue spaces of vegetation and water is good for health and wellbeing. This research, to date, mostly relates to green spaces (for example, Cleary et al., 2017; Wheeler et al., 2015), but increasingly researchers are extending this work into considerations of blue spaces (for example, Gascon et al., 2017).

The associated benefits of any form of bodily exercise for physical and mental health are documented through a biomedical frame drawing on biological and physiological understandings of how the body functions (Pinedo and Dahn, 2005; *https://www.nhs.uk/Livewell/fitness*). This includes psychological benefits to self-esteem and identity (Liu et al., 2015) in which gains for health are acquired and maintained for the individual body through individual physical exertion, achievement and sense of control. Through this lens, apart from variations in which muscles are exercised and the degree of exertion, swimming brings the same kinds of benefits as any other form of exercise and swimming outdoors may be little different to any other form of swimming. In a biomedical framing, while exercise may have pleasurable dimensions, these are only of interest for health gains in terms of furnishing the motivation to exercise. By contrast, framing benefits of exercise in terms of wellbeing, rather than a biomedical focus on physical and mental health, draws attention to the body as the site of emotions, sensations and social encounters. Explorations of these subjective experiences in natural environments identify therapeutic aspects of situation, immersion, symbolism and achievement which may all be integral to effecting higher self-esteem not only as mental states but also as embodied states (Bell et al., 2018; Doughty, 2013).

There is a suggestion in much of this research, intentionally or not, of universal principles at work. The Biophilia hypothesis, extended in this volume's focus on hydrophilia, explicitly argues that humans have an innate inclination to affiliate with other forms of life that is, at least in part, genetically evolved (Wilson, 1984). Other explicitly universalist explorations involve how humans respond to colours which, although a nascent science, suggests that the green-blue part of the spectrum may be relatively relaxing both psychologically and neurologically (Elliott and Maier, 2014); similar associations are proposed about the relaxing effects of the sounds of water (White et al., 2010). A distinctive strand of research with specific relevance to wild swimming argues for universal health and wellbeing benefits from regular immersion in cold water (Buckley, 2015). Less explicitly, research reporting a pattern of association between self-reported health and coastal proximity throughout England tends towards a universalistic tone (White et al., 2013). And while cultural meanings are often nodded to in passing, those special sites that have a reputation for healing and spiritual renewal are consistently, and across a range of different cultures, characterised by the presence of water which again hints at something essential and universal to human experience (Smyth, 2005). This tendency to imply that benefits from green and blue spaces are able to be generalised for all of humanity is countered in geographically informed work that explicitly foregrounds experience as always necessarily situated in given times and places. In this, the restorative effects claimed for being and moving in green and blue spaces are always necessarily produced relationally, culturally mediated and learned (Bell et al., 2018; Conradson, 2005). To date, much of the existing literature on the benefits of blue space relates to the experience of being near, beside or, occasionally, on water (Bell et al., 2015; Finlay et al., 2015; Nutsford et al., 2016; Roy, 2014; Völker and Kistemann, 2013). Swimming, however, adds a distinctive dimension to the experience of blue space through being fully in the water, of immersion. Physical immersion in the water might be seen as a direct corollary to the reported benefits to wellbeing that come from a sense of immersion in green space (Bell et al., 2018), but there is a significant difference in that water is not the given medium for human life. We do not routinely register the touch of being in air unless it is very windy; we may register temperature, but not touch. We are, however, very aware of the touch of being in water. Moreover, again in contrast to the immersive experiences of being in green spaces, we actively have to learn to be immersed in water, to swim, to breathe and to pay attention to the attendant risks (Straughan, 2012).

This distinctive quality of immersion in an 'alien' medium is associated with an emerging attention in both research and 'waterbiography' that privileges first-person experiential accounts of swimming. Research on the subjective benefits of being alongside green or blue space often underplays the presence of the material body in favour of cognitive accounts, but an account of swimming very much insists on the body's centrality. As its proponents will frequently point out, swimming, and especially wild swimming, exerts all parts of the body and involves awareness of all the senses, but with an unusual emphasis on the touch of the water and on proprioception, the sense of where parts of the body are in a context of feeling weightless or of flying in water (Straughan, 2012). The sensation of temperature of both the water and the body also features frequently in conversation among swimmers, at least in Northern climes, and constitutes one of the sources of risk to wild swimmers (Tipton and Bradford, 2014). Thus, Elizabeth Straughan describes her and others' experiences of scuba diving in Mexico through an emphasis on touch, on the sensuous encounter of the materialities of diving and the diving body with the water that produce the particular experience of place. She also attends to the sounds, the quietness of diving and how the practices of diving produce an almost meditative and calming encounter that chimes with the literature of therapeutic landscapes (Straughan, 2012). In a less exclusive type of swimming, Ronan Foley uses interviews with sea swimmers in Ireland for an account of how therapeutic benefits endure beyond the moment of the fleeting experience. The positive embodied emotions of Foley's swimmers occur repetitively and, as such, layer or sedi-ment a sense of being well accretively and in relation to the particular places and practices (Foley, 2017). At the same time, both studies flag the potential risks of wild swimming and the ways that swimmers monitor their environ-ment, or comment on irresponsible behaviour in others. Part of the gains for wellbeing comes through the acquisition of skills to function in the me-dium of water whether through training in swimming techniques, in using technologies as in scuba diving or in monitoring one's own body and the subsequent pleasure of exercising competency in this acquired expertise (see Straughan, 2012). Moreover, the heightened awareness of the body in the alien medium of water and the working of the body through swimming ef-fectively render the body differently as it becomes a body that feels at home in the water (Throsby, 2013).

## A dominant approach to wild swimming

The attention to the experience as encounter between water, body, sensa-tion and emotion is explicitly complex and relational; wellbeing is emergent within the assemblage of embodied material and emotional components. The academic work complements the highly embodied accounts in the emergent genre of waterbiography by theorising the wild swimming experi-ence. Nonetheless, despite the insightful contributions, this interacting set of autobiographical celebration and academic attention to embodied emo-tion presents wild swimming within a limited frame of understanding. First, despite the conceptual lens of relationality, the focus is sharply on a highly individualised and largely inward-looking experience. Indeed, the height-ened awareness of the body in water effectively emphasises the contours and borders of the individual self. Second, many of the activities explored, especially in the academic literature, require investment in equipment and technology and, as such, limit participation to higher income-bracket earn-ers. Sports England (2015) makes this observation about outdoor activities in general. The sea swimmers at two sites in Ireland in Foley's account of accretive wellbeing (2017) present a more inclusive form of wild swimming in that the sites are well established and normalised over a long history preceding the current trend, the activity is undertaken regularly and rou-tinely and the swimmers appear to require minimal kit. Elsewhere, however, including much of the United Kingdom, and despite a subgroup espousing a no-wetsuit approach, many of those wanting to participate in wild swim-ming all-year round will need as a minimum a full wetsuit. Beyond the sim-ple form of wild swimming, there is tendency in the academic literature to engage the higher skilled, technology supported water activities like scuba diving, marathon swimming and surfing, and often in exotic water tour-ism locations such as Mexico (see Straughan, 2012; Throsby, 2013). A third observation is that the academic literature is always very positive with a focus on the pleasures, the achievements and the benefits for wellbeing, an intellectualised variant of the more overtly promotional or waterbiographical writings by the avowed enthusiasts. Herein may lie a major problem with the existing accounts of wild swimming, both popular and academic, in that these are all accounts by people who like water, who like being in water and who like swimming. The term 'wild swimming' itself has gained currency from its high-profile use as the title of a book detailing places to swim outdoors in the United Kingdom (Rew, 2008). An account of experiencing wild swimming might look very different from a less convinced and committed informant group. Indeed, there is a substantial and very different body of literature that focuses on the attendant risks of wild swimming in terms of rip-tides (Brander, 2013), water temperature (Tipton and Bradford, 2014), pollution (Leonard et al., 2018) or associated sun exposure risks (Collins and Kearns, 2007). Moreover, even within the largely positively oriented therapeutic landscapes literature, there is recognition that benefits are relational, individual and may vary across time, especially the life-course (Finlay et al., 2015; Foley, 2015). Fourth, the existing literature is quite feminised, both in being largely written by women and in the sense of extolling more gentle benefits from the immersion. There are writings by men that are framed in a much more competitive voice in the language of overcoming adversity and in particular mastery over nature (for example, Walker, 2016). These other immersions are not entirely neglected in that more masculine immersions are often seen as the norm to be countered or at least complemented. Thus, Throsby's autobiographical account of marathon swimming explicitly counters the dominant narrative of achievement, mastery and overcoming adversity, gains felt only at the end of the swim, by describing benefits during the swim in terms of the feeling of the body's ease in the water (Throsby, 2013). Throsby engages gendered themes of the body further and extends her attention to how bodies feel differently competent in water to on land; bodies that feel disadvantaged on land either through physical impairment or through normative aesthetic biases come to feel themselves more positively in water. Finally, many of the accounts are very local and while the relationality and co-constitution of place, person, immersion and an emergent wellbeing offer important insights, wider contexts and connections are at best mentioned in passing and as such have only limited contribution to the question of why we are witnessing a marked growth in wild swimming now.

## Wider contexts of wild swimming

Whether framed as precognitive to cognitive or as mindfulness, stillness or getting in the flow, there is a hint in some of this work of trying to access an authentic experience, that is, an experience unmediated or constrained by language or interpretation. Moreover, the therapeutic benefits come in part from the opportunity to reconnect with oneself, to find or develop the authentic self through highly embodied activities such as swimming. In this, much of the writing on swimming resonates with the Romanticism of the early 1800s (Jarvis, 2015). Partly a reaction to the sweeping social and settlement changes of industrialisation and partly a reaction to the growth of scientific explanations of nature, the romantics privileged the free expression of creativity and emotions by the artist as untainted or constrained by outside influence. Nature countered the structures of the human world and a close connection with nature offered artistic health, a spur to the imagination and the opportunity to capture the ever-shifting inner emotional states. The romantic poets of 19th-century England are particularly closely associated with the outdoors, with the green and blue spaces of nature and several are known for participating in wild swimming, seeing it as an 'encounter with the sublime' (Parr, 2011). While Blake drew on nature to rail against the 'dark, satanic mills' in a new form of protest poetry, those following on, particularly Byron, Coleridge and Wordsworth, explicitly promoted nature as, on the one hand, healing and spiritually uplifting and, on the other, as a form of testing and development of the self through performance (Jarvis, 2015).

This suggestion of seeking an unmediated authentic experience to some extent takes a contemporary form through exhortations to find and grow an authentic inner self. This move to health and happiness as a result of working on oneself involves honing the body through exercise and diet and honing inner self-identity through positive thinking, learned optimism or mindfulness. This redirection of responsibility for wellbeing onto the inner self is evidenced not only in a health policy directed to individual actions, but also through the emergence of a new market in self-help books, courses and other commercial products (Davies, 2015; Ehrenreich, 2009). This inward-facing work is complemented by exhortations to outward-facing actions through joining groups, helping others or giving, as strategies to increase our social support and sense of self-worth (Nef, 2008). This opens all kinds of activities up as technologies of the self that are amenable to Foucauldian-inspired analyses of self-care, self-actualisation and social responsibility within a project of contemporary governmentality (Miller and Rose, 2007). Wild swimming, however, does not, of necessity, share romanticism's distrust of the processes of mass production, urban living and scientific explanation as alienating humanity from itself and its valorisation of an authentic humanity which can be restored through immersions in nature. A contemporary variant of this romantic ideology, as illustrated in much of the work on therapeutic landscapes (Smyth, 2005), promotes being in organic green and blue spaces for their positive benefits to individual interior human wellbeing. These impacts are often treated as accruing through a restorative process enabled through the meanings ascribed to a reconnection with nature. Such meanings may include an implicit assumption that, as humans, we all have an intrinsic need to feel connected with the other life forms of our planet. This differs from the markedly elitist Romantic notion of artistic authenticity of experience and expression in being a sensibility that is shared across humanity. This need for connection is reflected by Robert Macfarlane, another major writer in the new-nature writing genre, who also serves as Patron for the Outdoor Swimming Society. Promoting their work in 2008, he wrote:

'over the past decade or so, a desire for what might be termed "reconnection" has emerged, a yearning to recover a sense of how the natural world smells, tastes and sounds. More and more people are being drawn back to the woods, hills and waters of Britain and Ireland' (https://www. outdoorswimmingsociety.com/patron-statement/).

McFarlane's desired reconnection to the natural world sees us 'drawn back' to a connection that we are 'yearning to recover' having been lost to us through modern ways of living. Research has mirrored this orientation, including, for example, through psychological explanations of the benefits of being in nature that focus on an increased sense of connectedness to nature, improved attentional capacity, greater positive emotion and better ability to reflect on life (Mayer et al., 2009). In this mode of understanding, the reconnection to the natural world restores our sense of wellbeing, complementing rather than reproducing the Romantics belief in the need to connect with nature to escape the intrusions and influence of modern living on authentic self-expression. In the literature on therapeutic landscapes, a similar use of the language of recovery and restoration (Conradson, 2005; Korpela et al., 2010) begs the question of recovery from what and in this framing, it is modern living itself that is too demanding (Carlisle et al., 2012), and from which we all need to retreat periodically, often into green or blue, spaces and their restorative potential for being well. Nonetheless, despite evident differences of focus, these two forms of valuing a connection to nature also infuse each other, in that the Romantic imaginary itself is part of the relational process through which a nature-person encounter becomes constituted as therapeutic (Conradson, 2005).

This move to self-actualisation appears, then, to share with Romanticism a central focus on the authentic individual whether as individual self-actualisation or as individualised self-expression. In a society promoting exercise and where leisure itself is highly managed and highly commoditised, including the use of some green and blue spaces (for example, Conradson, 2007; Foley et al., 2011), expressing individuality becomes ever more challenging; many outdoors activities, including the Great Swim events, are becoming annual mass events, highly managed and involving fundraising for charities. The growth of wild swimming has then, perhaps inevitably as its popularity has grown, manifest two coexistent trends: the commoditised leisure for self-care and social contribution through the large events of mass swims in which participants typically raise money for charity, and the solitary or small group swims in an ever-expanding range of remote or 'wild' locations. For some, sociality is the main appeal, but for others there is a constant search for new experience and new expressions of individuality.

The research on the benefits of immersion in nature suggestive of this coming together of exhortations to self-care and the romantic quest for self-expression mostly ignores considerations of class. The limited work available clearly indicates the association of class with the extent of engagement with nature (Bell et al., 2017; Korpela et al., 2010), suggesting that access to blue space should similarly be seen as an issue of environmental justice (Raymond et al., 2016). By contrast, wild swimming is often promoted by its enthusiasts as the ultimately inclusive form of physical activity, offering spaces of participation that cut across differentiated communities, generations and social categories. Most people in the United Kingdom now learn to swim at school and wild swimming requires no equipment except, usually, a swimsuit. This claim for wild swimming as a democratic and egalitarian activity does appear to deliver with respect to gender. A notable characteristic of the growth in the numbers registered to join the organised wild swimming organised events has been the wide appeal of these and especially to older and female participants. Just over half of the participants in the 2014 10-km marathon swim in Lake Windermere were female, making wild swimming unusual among outdoor activities in having a higher participation of women than men (Laville, 2015). The contemporary attraction of swimming for women builds on a history of what is perhaps a surprising acceptability of female participation in wild swimming. Therapeutic bathing by women at the seaside, for example, was acceptable, even encouraged, from 18th-century Britain onwards as long as women's bodies were visually secluded by entering the water from a bathing machine (Jarvis, 2015; Parr, 2011). In the late 1800s, demonstrations of swimming and outdoor long-distance swimming gained popularity as a public spectacle following the first swim by Matthew Webb in 1875 across the Channel between England and France. And this was a spectacle that included and celebrated participation by women. One example was Agnes Beckwith, a performance swimmer in the late 1800s, who was one of a number of women who displayed feats of swimming in a tank publicised as 'The Aquarium', and performed long-distance swims and races as public spectacle (Day, 2012). While the participation of women in such performances of athleticism and swimming skills is in some ways surprising, water has a long history of symbolic association with the feminine, at least until water became more explicitly a commodity (Strang, 2014).

The history of bathing more generally documents its uptake as a fashionable but exclusive activity primarily for therapeutic purposes in the 18th century among the wealthy (Sutherland, 1997); as bathing became seen as a pleasurable, leisure activity, the development of railways in 19th-century Victorian Britain enabled a more democratic participation that was accompanied not only by the rise of commercialised popular seaside resorts but also by a range of class-based conflicts, centred on bathing, conduct and norms of respectability (Parr, 2011). From the 18th century onwards, resorts slightly further afield from the large cities started to become more attractive to the well-to-do (see Sutherland, 1997). An insistence that we need to include considerations of class within understandings of contemporary wild swimming may expose ways in which connections with green and blue spaces express a distinctly middle-class citizenship while simultaneously enabling a moral judgement of others.

Contemporary locations for wild swimming continue not to be equally accessible for all, not only through physical distance and financial constraints but also through social barriers. Even while officially open to all, places can generate a strong sense of who belongs in them and who does not. This exclusion is subtle, grounded often in cultural resonances and meanings for different population groups through historical use and familiarity. Such historical resonance is likely affirmed through the presence and sometimes behaviour of those who do routinely use a particular space as explored in a case of the experiences of discomfort in inhabiting a public beach in New England, USA, conventionally used by affluent residents of the neighbourhood (Keul, 2015). The sense of exclusion engendered on the New England beach illustrates the ways in which outdoor places and activities constitute particular social spaces, imbued with norms and shaping given practices. Drawing on Bourdieu, taking up a new activity, such as wild swimming, may reflect the extent of congruence between the norms of the particular social space with the potential participants' internalised sets of values and perceptions (Nettleton and Green, 2014). Those feeling comfortable in alternative spaces of activity such as wild swimming are those who already share this situated and taken-for-granted, or tacit, knowledge of practice. Moreover, they not only feel they belong, but can become self-congratulatory about their activity choices and moralistic towards others judged as lacking the cultural capacities to make decisions for their wellbeing (Guthman, 2003). Leila Dawney draws on embodied and social imaginaries to comprehend how we think, feel and practise being in nature reflects ours and others previous encounters in the world (Dawney, 2011). An embodied imagination is central to understanding how some feel a close sense of affiliation or belonging while others may feel a powerful sense of alienation or exclusion.

> A focus on practices as productive of particular landscape imaginaries enables a consideration of how these imaginaries play out through bodies and take material form, producing certain kinds of engagements and precluding others, forming connections and disjunctures between bodies and offering experiences variously felt by different subjects.
>
> (Dawney, 2014: 90)

This approach allows the reflective subject to weave together the diverse and sometimes conflicting imaginaries of Romanticism, of alienation and reconnection to nature, of self-care and self-actualisation but further attends to the contouring of experience that is differentiated by social identities such as class, age, gender, ethnicity, mobility and so forth. Moreover, a contemporary imaginary of Romanticism as modern life alienated from an authentically human connection with nature may intersect with a noted tendency for working class lives to be stereotypically depicted, both by others and themselves, as closer to some imaginary of an authentic mode of existence. This generates an imaginary that may inform a particularly middle-class anxiety about the lack of connection to the natural world and to each other which becomes expressed through choices to participate in evidently counter-modern leisure practices that are closer to nature (Dawney, 2014).

## Swimming and the wild, blue 21st-century citizen

The growth of the modern passion for wild swimming can be understood as emerging from an interweaving of several narratives about our relationship to nature. These include an inherent need for human connection to nature, an alienation from ourselves through modern living, the need to retreat and recover from the demands of modern living and the good citizen as one taking care of themselves and their own wellbeing. While many of these reconfigure pre-existing imaginaries, this reconfiguration is made through the contemporary lens of wellbeing together with a narrative for the individualised responsibility for the embodied self, both of which are characteristic of modern forms of liberal governance. Research that intimates the ways in which the potential benefits of new technologies of self-care, such as wild swimming, may be differently distributed by age, class, gender, ethnicity and so forth redirects our attention back to the subtleties of how experience is itself systematically differentiated. Progressing an account of wild swimming that attends to social justice demands greater attention to how wild swimming itself is differentiated in terms of locations, organisational structures, technologies of immersion and socio-demographics of participation. While this differentiation of locations that are variously more popular or more elitist in access is recognised for swimming bath, lidos, spas and so forth, this is less obviously visible and under-researched in relation to outdoor settings. Foley's processes of accretive wellbeing, Bourdieu's notions of habitus, field and tacit knowledge, and Dawney's embodied imaginaries provide conceptual tools with rich potential for advancing our understanding of the benefits and limitations of wild swimming to engage nature and enhancing wellbeing in particular times and places.

The focus within the attention to the immersive experiences of wellbeing in green and blue spaces not only ignores class, but also more surprisingly side-steps direct engagement with environmental issues. These accounts, both intellectual and autobiographical, all reaffirm, even in those attending to the social, the dominant privileging of individual subjectivity within contemporary approaches to understanding our encounters with nature. Regardless of the explanatory frames drawn on, both waterbiographies and academic interrogations of wild swimming reflect back ultimately to an individualised purpose, whether that of quest, retreat or connection. While the experiential accounts often attend in tones of wonder to the aesthetics and sensory pleasures of nature, these primarily nurture human experience rather than draw human attention to the natural world. Even the activism of the Outdoor Swimming Society primarily targets increasing public access to wild swimming locations with concerns over pollution directed to the protection of wild swimmers. The writer Robert McFarlane argues that the British 'have long specialised in a disconnect between their nature romance and their behaviour as consumers'. He continues with his theme by arguing that 'British parochialism – its strong tradition of interest in the local – leads too often only to general conclusions: to a comfortable sentimentalism' (McFarlane, 2005). While the accounts of taking up and practising wild swimming are often anything but 'comfortable' in their bodily experiences, McFarlane's claim for sentimentalism in relation to our encounters with nature in general and water in particular and a disconnect between our local perception of nature and any wider engagement with environmental concerns deserve our serious attention in future interrogations of our practices as wild, blue 21st-century citizens.

## Acknowledgements

The development of the chapter was supported through a Wellcome Trust discretionary award: WT 209513.

## References

Bell, S.L., Phoenix, C., Lovell, R. and Wheeler, B.W. (2015). Seeking everyday well-being: The coast as a therapeutic landscape. *Social Science and Medicine,* 142, 56–67.

Bell, S.L., Westley, M., Lovell, R. and Wheeler, B.W. (2018). Everyday green space and experienced well-being: The significance of wildlife encounters. *Landscape Research,* 43, 8–19.

Bell, S.L., Wheeler, B.W. and Phoenix, C. (2017). Using geonarratives to explore the diverse temporalities of therapeutic landscapes: Perspectives from 'green' and 'blue' settings. *Annals of the American Association of Geographers,* 107, 93–108.

Bolz, D. (2012). Creating places for sport in inter-war Europe: A comparison of the provision of sports venues in Italy, Germany and England. *The International Journal of the History of Sport,* 29, 1998–2012.

Brander, R.W. (2013). Thinking space: Can a synthesis of geography save lives in the surf? *Australian Geographer,* 44, 123–127.

Buckley, J.J. (2015). *Cool Swimming.* London, South London Swimming Club.

Carlisle, S., Hanlon, P., Reilly, D., Lyon, A. and Henderson, G. (2012). Is 'modern culture' bad for our wellbeing? Views from 'elite' and 'excluded' Scotland. In, S. Atkinson, Fuller, S. and Painter, J. (eds), *Wellbeing and Place.* Farnham, Ashgate, 123–140.

Cleary, A., Fielding, K.S., Bell, S.L., Murray, Z. and Roiko, A. (2017). Exploring potential mechanisms involved in the relationship between eudaimonic wellbeing and nature connection. *Landscape and Urban Planning,* 158, 119–128.

Collins, D. and Kearns, R. (2007). Ambiguous landscapes: Sun, risk and recreation on New Zealand beaches. In, A. Williams (ed), *Therapeutic Landscapes.* Farnham, Ashgate, 15–32.

Conradson, D. (2005). Landscape, care and the relational self: Therapeutic encounters in rural England. *Health and Place,* 11, 337–348.

Conradson, D. (2007). The experiential economy of stillness: Places of retreat in contemporary Britain. In, A. Williams (ed), *Therapeutic Landscapes.* Farnham, Ashgate, 33–48.

Davies, W. (2015). *The Happiness Industry.* London, Verso.

Dawney, L. (2011). Social imaginaries and therapeutic self-work: The ethics of the embodied imagination. *The Sociological Review,* 59, 535–552.

Dawney, L. (2014). 'Feeling connected': Practising nature, nation and class through coastal walking. In, P. Gilchrist, Carter, T. and Burdsey, D. (eds), *Coastal Cultures: Liminality and Leisure.* Eastbourne, Leisure Studies Association, 89–103.

Day, D. (2012). 'What girl will now remain ignorant of swimming?' Agnes Beckwith, aquatic entertainer and Victorian role model. *Women's History Review,* 21, 419–446.

Deakin, R. (1999). *Waterlog: A Swimmer's Journey through Britain.* London, Vintage Books.

Dogliani, P. (2000). Sport and Fascism. *Journal of Modern Italian Studies,* 5, 326–348.

Doughty, K. (2013). Walking together: The embodied and mobile production of a therapeutic landscape. *Health and Place,* 24, 140–146.

Ehrenreich, B. (2009). *Smile or Die: How Positive Thinking Fooled America and the World.* London, Granta.

Elliott, A.J. and Maier, M.A. (2014). Colour psychology: Effects of perceiving color on psychological functioning in humans. *Annual Review of Psychology,* 65, 95–120.

Etherington, J. (2015). Why open-water swimming is the trend of 2015. *The Telegraph* 5th January. Available at: http://www.telegraph.co.uk/comment/personal-view/11325401/Why-open-water-swimming-is-the-trend-of-2015.html, last accessed January 8, 2015.

Finlay, J., Franke, T., McKay, H. and Sims-Gould, J. (2015). Therapeutic landscapes and wellbeing in later life: Impacts of blue and green spaces for older adults. *Health & Place,* 34, 97–106.

Fitzmaurice, R. (2017). *I Found My Tribe.* London, Chatto and Windus.

Foley, R. (2015). Swimming in Ireland: Immersions in therapeutic blue space. *Health & Place,* 35, 218–225.

Foley, R. (2017). Swimming as an accretive practice in healthy blue space. *Emotion, Space and Society,* 22, 43–51.

Foley, R., Wheeler, A. and Kearns, R. (2011). Selling the colonial spa town: The contested therapeutic landscapes of Lisdoonvarna and Te Aroha. *Irish Geography,* 44, 151–172.

Gascon, M., Zijlema, W., Vert, C., White, M.P. and Nieuwhuijsen, M.J. (2017). Outdoor blue spaces, human health and wellbeing: A systematic review of quantitative studies. *International Journal of Hygiene and Environmental Health,* 220, 1207–1221.

Guthman, J. (2003). Fast food/organic food: Reflexive tastes and the making of 'yuppie chow'. *Social and Cultural Geography,* 4, 45–58.

Heminsley, A. (2017). *Leap In.* London, Windmill Books.

Jarvis, R. (2015). Hydromania: Perspectives on romantic swimming. *Romanticism,* 21, 250–264.

Keul, A. (2015). The fantasy of access: Neoliberal ordering of a public beach. *Political Geography,* 48, 49–59.

Korpela, K.M, Ylén, M., Tyrväinen, L. and Silvennoinen, H. (2010). Favorite green, waterside and urban environments, restorative experiences and perceived health in Finland. *Health Promotion International*, 25, 200–209.

Landreth, J. (2017). *Swell: A Waterbiography*. London, Bloomsbury.

Laville, S. (2015). Different Strokes: Open-Water Swimming Takes the UK by Storm. *The Guardian*, 18 August. Available at: https://www.theguardian.com/lifeandstyle/2015/aug/18/different-strokes-open-water-swimming-uk-event-lake-river-seal, last accessed August 18, 2017.

Lee, J.J. (2017). *Turning: A Swimming Memoir*. London, Virago Press.

Leonard, A.F.C., Zhang, L., Balfour, A.J., Garside, R., Hawkey, P.M., Murray, A.K., Ukoumunne, O.C. and Gaze, W.H. (2018). Exposure to and colonisation by antibiotic-resistant *E. coli* in UK coastal water users: Environmental surveillance, exposure assessment, and epidemiological study (Beach Bum Survey). *Environment International*, 114, 326–333.

Liu, M., Wu, L. and Ming, Q. (2015). How does physical activity intervention improve self-esteem and self-concept in children and adolescents? Evidence from a meta-analysis. *Plos One*, 10, e0134804.

Marino, G. (2010). The emergence of municipal baths: Hygiene, war and recreation in the development of swimming facilities. *Industrial Archaeological Review*, 32, 35–45.

Mayer, F.S., Franatz, C.M., Bruehlman-Senecal, E. and Dolliver, K. (2009). Why is nature beneficial? The role of connectedness to nature. *Environment and Behaviour*, 41, 607–643.

McFarlane, R. (2005). Where the Wild Things Were. *The Guardian*. Available at: https://www.theguardian.com/books/2005/jul/30/featuresreviews.guardianreview22, last accessed January 26, 2018.

Miller, P. and Rose, N. (2008). *Governing the Present*. Cambridge, Polity Press.

Minihane, J. (2017). *Floating: A Life Regained*. London, Duckworth Overlook.

Moran, J. (2014). A cultural history of the new nature writing. *Literature and History*, 23, 49–63.

New Economics Foundation. (2008). *Five Ways to Wellbeing*. London, New Economics Foundation. Available at: https://neweconomics.org/uploads/files/d80eba95560c09605d_uzm6b1n6a.pdf [accessed 30.11.18]

Nettleton, S. and Green, J. (2014). Thinking about changing mobility practices: How a social practice approach can help. *Sociology of Health and Illness*, 36, 239–251.

Nutsford, D., Pearson, A., Kingham, S. and Reitsma, F. (2016). Residential exposure to visible blue space (but not green space) associated with lower psychological distress in a capital city. *Health & Place*, 39, 70–78.

Parr, S. (2011). *The Story of Swimming: A Social History of Bathing in Britain*. Stockport, Dewi Lewis Publishing.

Peters, A.F. (2014). *Dip: Wild Swims from the Borderlands*. London, Rider Books.

Pinedo, F.J. and Dahn, J.R. (2005). Exercise and well-being: A review of mental and physical health benefits association with physical activity. *Current Opinion in Psychiatry*, 18, 189–193.

Pussard, H. (2007). Historicising the spaces of leisure: Open air swimming and the lido movement in England. *World Leisure Journal*, 49, 178–188.

Raymond, C., Gottwald, S., Kuoppa, J. and Kyttä, M. (2016). Integrating multiple elements of environmental justice into urban blue space planning using public participation geographic information systems. *Landscape and Urban Planning*, 153, 160–169.

Rew, K. (2008). *Wild Swim*. London, Guardian Books.

Rhodes, D. (2016). Swimming participation in England falls by nearly 24%. *The BBC*. Available at: https://www.bbc.co.uk/news/uk-england-38128659, last accessed July 19, 2018.

Roy, G. (2014). Taking emotions seriously: Feeling female and becoming-surfer through UK surf space. *Emotion, Space and Society*, 12, 41–48.

Shadrick, T. (2017). *Watermarks: Writing by Lido Lovers and Wild Swimmers*. Folkestone, Frogmore Press.

Smyth, F.S. (2005). Medical geography: Therapeutic places, spaces and networks. *Progress in Human Geography*, 29, 488–495.

Sports England. (2015). *Getting Active Outdoors: A study of Demography, Motivation, Participation and Provision in Outdoor Sport and Recreation in England*. London, Sports England.

Strang, V. (2014). Lording it over the goddess: Water, gender and human-environment relations. *Journal of Feminist Studies in Religion*, 30, 85–109.

Straughan, E.R. (2012). Touched by water: The body in scuba diving. *Emotion, Space and Society*, 5, 19–26.

Sutherland, E. (1997). "A little sea-bathing would set me up forever" the history and development of the English seaside resorts. *Persuasions*, 19, 60–76.

Throsby, K. (2013). 'If I go in like a cranky sea lion, I come out like a smiling dolphin': Marathon swimming and the unexpected pleasures of being a body in water. *Feminist Review*, 103, 5–22.

Tipton, M. and Bradford, C. (2014). Moving in extreme environments: Open water swimming in cold and warm water. *Extreme Physiology and Medicine*, 3, 12.

Völker, S. and Kistemann, T. (2013). "I'm almost entirely happy when I'm here!" Urban blue enhancing human health and well-being in Cologne and Dusseldorf, Germany. *Social Science and Medicine*, 78, 113–124.

Walker, A. (2016). *Man vs Ocean: One Man's Journey to Swim the Seven Seas*. London, John Blake Books.

Wardley, T. (2017). *The Mindful Art of Wild Swimming: Reflections for Zen Seekers*. Brighton, Leaping Hare Press.

Wheeler, B.W., Lovell, R., Higgins, S.L., White, M.P., Alcock, I., Osborne, N.J., Husk, K., Sabel, C.E. and Depledge, M.H. (2015). Beyond green space: An ecological study of population general health and indicators of natural environment type and quality. *International Journal of Health Geographics*, 14, 17. doi: 10.1186/s12942-015-0009-5.

White, M.P., Alcock, I., Wheeler, B.W. and Depledge, M.H. (2013). Coastal proximity, health and well-being: Results from a longitudinal panel survey. *Health and Place*, 23, 97–103.

White, M., Smith, A., Humphreys, K., Pahl, S., Snelling, D. and Depledge, M. (2010). Blue space: The importance of water for preference, affect and restorativeness ratings of natural and built scenes. *Journal of Environmental Psychology*, 30, 482–493.

Wilson, E.O. (1984). *Biophilia*. Boston, MA, Harvard University Press.

# 13 Environmental uncertainty and muddy blue spaces

## Health, history and wetland geographies in Aotearoa New Zealand

*Meg Parsons*

## Introduction

Since 1840, when the British colonisation of Aotearoa New Zealand formally commenced with the signing of the Treaty of Waitangi, there has been a significant reduction in the total area and health of wetlands (more than 90% loss) as a result of drainage and river 'improvement' schemes (Ausseil et al., 2015; Clarkson et al., 2013). These figures are among the greatest extent of wetland reduction in the developed world (Clarkson et al., 2013). The coupled social-ecological transformation of wetlands to grasslands and urban areas represents a familiar modernising development pathway adopted in different settler colonial societies, yet one which is often overlooked by scholars, natural resource managers and members of the public. Indeed, the status of land and water (its availability, ownership, quality, management, degradation) receives far greater attention (Beattie and Morgan, 2017; Parsons and Nalau, 2016: Romero Lankao, 2010). Yet, as I will outline in this chapter, wetlands (the muddy blue spaces of brown and green, in-between hybrid spaces between fixed land and fluid water) were and are fundamental components of healthy ecosystems interconnected to the health and wellbeing of both human and non-human communities.

When thinking about current environmental crises and future impacts of climate change, it is critical that we contemplate how certain human-environment relations (specifically perceptions, values and practices around what constitutes healthy blue spaces) are institutionalised in dominant resource management policies. A historical perspective of how people (as individuals and members of wider social groups) perceived, interacted with and related to specific freshwater spaces in the past provides importance insights into current and future hydro-social cycles, path dependencies and the multiplicity of healthy waters. In this chapter, I provide an examination of drainage of the wetlands and reconfiguration of floodplains of Aotearoa New Zealand, focusing specifically on the Rangitāiki wetlands during 1910s–1940s. I document how the emergent settler colonial government and individual white settlers articulated and acted on specific geographical imaginations that determined what landscapes and waterscapes should look like and how those spaces interacted with. Interventions (legal, engineered and discursive) to define and reform wetlands (muddy blue spaces, liminal zones) into fixed straight blue lines running through neatly ordered green pastures filled with certain fauna (cows, sheep) involved the expression of particular values about what environments were deemed desirable, which were built on particular understandings of health and 'nature'. These changes were reflective of different values Māori and European settlers (Pākehā) attached to the wetland blue spaces, and resulted in the disruption of pre-existing relationships of Māori communities with local wetlands, the diminishment of communities' hydro-resilience.

In Aotearoa New Zealand there are a variety of wetlands and associated ecosystems. These muddy blue spaces are spaces of transitions between waters and lands. Throughout the 19th and 20th centuries in Aoteaora New Zealand, the term 'swamp' was used in a multitude of settings by Pākehā to refer to any area with pooled water and some associated vegetation cover. Technically, however, the term 'swamp' now is used only in reference to wetlands that are forested with wooded vegetation (Park, 2001: 26). Scientists classify wetlands in different ways, such as palustrine (emergent plants over freshwater), estuarine (estuaries and lagoons), lacustrine (lakes and ponds), riverine, interior and coastal, or swamps, bogs and mires (Clarkson et al., 2013). Mātauaranga Māori (Māori knowledge) uses different terms of classification for wetlands, including poharu (palustrine), roto and moana (lascustrine), awa and manga (riverine), and wahapū, hāpua and muriwai (estuarine). All major types of wetlands are found in Aotearoa New Zealand.

In this chapter, I draw on environmental history, social history of medicine, cultural geography, indigenous geography and global environmental change scholarships to weave together a narrative about how Māori and Pākehā imagined and interacted with the wetlands of the Aotearoa New Zealand in the 19th and first half of the 20th centuries. Analysis centres on government reports, photographs and newspaper articles, as well as other documents, held by the National Archives of New Zealand as well as published memoirs and oral history collections. In particular, I concentrated on materials pertaining to the Rangitāiki Plains, which was formerly wetlands and encompasses the towns of Whakatāne, Matatā, Te Teko and Edgecumbe, and is the area where I spent a great deal of time as a child and adult. Through these documents we can see glimpses of larger changes in the ways wetlands, rivers and floods were understood before and after drainage works were undertaken in the 1910s–1940s.

In this chapter, I first outline historical and geographical scholarship examining anxieties about environmental conditions in the 19th century. I then proceed to discuss how various medical theories of disease-causation contributed to white perceptions of wetlands as unhealthy spaces throughout the 19th and early 20th centuries. Finally, I analyse the drainage of the Rangitāiki wetlands between the 1910s and 1940s, and demonstrate how hydrological interventions contributed to the loss of resilience and increased vulnerability to flooding.

## Environmental health and settler anxieties in the 19th-century Aotearoa New Zealand

From the outset of European encounters with the wetlands of Aotearoa New Zealand, European explorers, missionaries and settlers perceived the muddy fluid blue spaces as deeply problematic 'liminal zones'. Far from being therapeutic spaces, as perceived by Māori communities, living in or visiting wetlands was deemed by European settlers to be both 'ruinous to health' (the cause of fevers, rheumatism and other illnesses) and 'retarding settlement' (Anonymous, 1879). In 1875, for instance, the outbreak of fever in the township of Napier (Hawke's Bay district) was blamed on the wetlands that surrounded the town and the colony's first drainage legislation was introduced as a result of public health fears (New Zealand Parliament, 1875). Such environmental anxieties and the links between colonial environments, climates, biota and health featured prominently in public and political discussions in other colonial societies, including Australia, Canada and South Africa, and prompted similar attempts to control and re-engineer land, waters and biota (Beattie, 2005; Bonnell, 2014; Flikke, 2016; Giblett, 2014; Webster and Mullins, 2003). Such environmental anxieties were reflections of colonial actors' perceptions of their own vulnerabilities in new landscapes and waterscapes (Beattie, 2008; Morgan, 2013; Park, 2002).

Popular medical and geographical discourses of the time period framed wetlands as unhealthy and unproductive spaces that required efforts to remake them (Beattie, 2005; Park, 2002). Miasmatic thinking, a popular theory of disease transmission throughout European societies until the mid- to late 19th century, understood ill health as a result of exposure to miasma (bad airs), which could be generated from rotting vegetation and animal matter, human waste, stagnant water and foul air from factories, people and so forth (Bashford, 1998; Beattie, 2008). Wetland abundance of decomposing plants and torpid murky ponds was viewed and smelt (with olfactory sense fundamental) as the home of countless dangerous miasma. However, the connection between causation and transmission of diseases was not clearly defined nor indeed was this linkage viewed as being of particular importance. Rather emphasis was placed on the regulation of indoor and outdoor spaces, with wetland drainage, removal of indigenous vegetation and the planting of certain healthy (non-indigenous) trees (oaks, eucalyptus, willows, pine) all deemed to be necessarily for improving the health of (white) bodies and modernising (and civilising) landscapes (Beattie, 2005, 2008). The drainage of Rangitāiki wetlands in the first half of the 20th century, which I will now discuss, is a prime example of how state-directed colonial hydrological interventions involved the consolidation of the power of the settler colonial state, the creation of new settler-dominated grasslands' economies and landscapes, the marginalisation of Māori communities and the diminishment of 'hydro-resilience' (Beattie and Morgan, 2017).

### 'Unwatering' and remaking muddy blue spaces of the Rangitāiki wetlands

The Rangitāiki wetlands, located in the Eastern Bay of Plenty in Aotearoa New Zealand's North Island, traditionally comprised the area of 32,000 acres including the Tarawera River (western boundary), the Whakatāne River (eastern boundary) and the Rangitāiki River (which ran through the middle) (Parsons and Nalau, 2016) (see Figure 13.1). All three rivers left the mountain ranges as separate rivers but converged into a single wetlands area, which formed a number of lagoons. The wetlands drained into the ocean through three main channels at Orini, Awaiti and Mātatā; however, the channels frequently shifted with flood events, storm surges, coastal erosion and shifting sand dunes making for highly dynamic fluid waterscapes/landscapes.

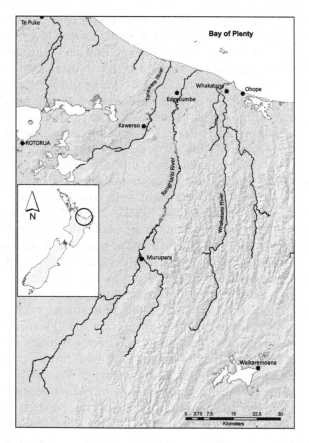

*Figure 13.1* Map of location of Rangitāiki Plains and present-day course of the three rivers.

Source: Author.

For Māori the Rangitāiki wetlands, with its three awa (rivers), springs, muddy spaces and biota, were (and are still) central rather than liminal zones, both materially (as sites of cultivations, food harvesting and healing) and metaphysically (as places of spiritual meaning) significant to Māori health and wellbeing (Rangitaiki River Forum, 2015; Waitangi Tribunal, 1999). Archaeologists argue that unlike in European cultures where wetlands were marginal places, in Aotearoa New Zealand wetlands were central to Māori culture. Archaeological evidence identifies the existence of wetland pā (fortified villages) within the area, with artefacts found including buildings, canoe and horticultural tools, highlighting the continuity of occupation in the area (Horrocks et al., 2004; Irwin, 2013). In 1840, when British colonial rule formally commenced in Aotearoa, numerous different Māori iwi (tribe) and hapū (subtribe) lived within or near the three river catchments and used the resources of the Rangitāiki wetlands, with usage rights overlapping between different hapū. This includes hapū of Ngāti Awa, Ngāti Tūwharetoa, Ngāti Makino, Ngāti Pikiao, Ngāti Manawa and Tūhoe (Rangitaiki River Forum, 2015). The wetlands were crucial resource gathering areas for Māori. A variety of plant species were harvested by Māori; for instance, harakeke (New Zealand flax – *Phormium tenax*) was used for clothing, mats, baskets and rope; kuta (bamboo spike sedge – *Eleochiris sphacelata*) for insulation and weaving; and raupō (*Typha orientalis*) for food and thatching. Similarly the Rangitāiki wetlands provided habitat for tuna (eels – *Anguilla spp.*), inanga (whitebait – *Galazias spp.*) and other fish species which were important food sources for Māori, as well as numerous bird species (Clarkson et al., 2013; Park, 2001; Parsons and Nalau, 2016). The higher areas of land were used for cultivating crops (first kumara and taro, and later with the arrival of Europeans, potatoes and wheat, among others).

In Māori ontologies, social and ecological systems are indivisible wholes (Panelli and Tipa, 2007; Salmond, 2017). Unlike Western scientific traditions, no clear division made between separate elements of freshwater systems, such as the water, riparian, river bed, wetlands or estuaries. From Māori world view, wetlands are perceived as being interconnected with other water sources, including rivers, streams and springs, and estuaries. Indeed, prior to the European colonisation in the plains of Aotearoa New Zealand (as the Rangitāiki wetlands example demonstrates), the majority of rivers and streams ran through areas of vast wetlands, and thus rivers were indistinguishable from wetlands (Clarkson et al., 2013; Denyer and Robertson, 2016).

The metaphysical and material aspects of rivers, wetlands and other blue spaces are woven together, and linked to the wellbeing of the local Māori. Māori conceptualisations of wellbeing, unlike those of Western scientific notions of individuals' physical and mental health status, centre on the health of the self, whanau (extended family), hapū (subtribe) and iwi (tribe) in connection with their traditional lands and waterways (rohe) (Panelli and Tipa, 2007); emphasis is placed on the interrelationships between individuals and collectives (human, ecological and metaphysical communities) and on the responsibility of local Māori (tangata whenua) to maintain the mauri (life force) of their rohe (Rangitaiki River Forum, 2015; Te Pahiopoto Hapu, 2017). For whānau and hapū the muddy blue spaces of the Rangitāiki wetlands and three rivers were, thus, intertwined with their physical and spiritual health, providing basic provisions (food, shelter, water), as well as sites of healing and connections with the spiritual worlds. Waka (canoes) were used to traverse the wetlands and rivers, for both transport and trading. Whānau and hapū maintained customary usage rights to the resources and usage of particular places including tauranga ika (landing places for canoes), pa tuna (eel weirs) and fishing locations. Ngāti Manawa and Ngāti Whare, whose rohe (traditional lands and waters) is located in the middle and upper reaches of the Rangitāiki River, described their 'eel culture' prior to their dispossession and the environmental degradation of the area. Elders report how tuna and tuna harvesting, and 'the rivers have always been the lifeblood of the people'. Different varieties of eels and other fish species were identified as living in specific areas. 'In different places, they tasted different; some were ordinary, some were special, but all were considered taonga' (meaning sacred) (Waitangi Tribunal, 1998: 12). In addition to food harvesting, the waters of the Rangitāiki River were used in healing and spiritual practices. Family members (whanau) ensured that those living outside the rohe were supplied with the water of the Rangitāiki River for special occasions. As an elder recounted to the Waitangi Tribunal,

> The water from the puna wai [water of the spring] of a whanau is considered to a taonga [a treasure] to that whanau as it carries the Mauri [life force] of that particular whanau [family]. Of course all the waters of the puna wai find their way into the river and thereby join with the Mauri of the river. In essence then the very spiritual being of every whanau is part of the river ... In this sense the river is more than a taonga[;] it is the people themselves.
>
> (Waitangi Tribunal, 1998: 13)

Wetlands and rivers are home to both the physical and the metaphysical (Te Aho, 2014). Taniwha, supernatural beings who could be either friendly or hostile and often assume eel-like forms, dwell in the murky waters of awa and linked to specific hapū and iwi (Dodd, 2010; Kolig, 2007; Rangitaiki River Forum, 2015; Salmond, 2017). Blue spaces therefore were (and are) multiple things, assemblages, connections and relationships, with rivers the embodiment of ancestral relationships through genealogical connections (whakapapa) to specific whanau, hapu and iwi, and wetlands places of home, healing, physical and spiritual healing. Rivers and wetlands all said to possess a mauri, which Māori were required to maintain through correct behaviours and practices. In these complex fluid and interconnected landscapes-waterscapes, physical features were and are more than simply wetlands, rivers, streams, waterfalls and wetlands; they carry with them histories of past events, stories and narratives that give meaning to places and link them with whakapapa (Parsons et al., 2017; Te Aho, 2014). Thus, the alteration of these places of meaning as a result of settler-led colonial policies and interventions resulted in substantive losses (of access to resources, of places of recreation, of economic activities and of spiritual connection) for Māori communities, including diminishing health and wellbeing.

Like elsewhere in the Aotearoa New Zealand, Māori used the geothermal springs (ngāwha, puia) located on the eastern and central edge of the Rangitāiki wetlands for cooking, bathing, laundry, recreation, healing and spiritual purposes. Many geothermal areas were important wahi tapu (sacred spaces) and relationships between people and places were active, ongoing and multiple (Foley et al., 2011; Rockel, 1986; Stokes, 2000). While Europeans similarly valued hot springs, such interactions chiefly centred on visiting hot springs (at spa towns) for health purposes rather other purposes. In the Rangitāiki, the Māori frequent use of these hot blue spaces was criticised by government doctor (Matthew Scott) in 1885 for being not only over-indulgent but also unhealthy. Māori living near the geothermal springs at Awakeri supposedly suffering from high rates of tuberculosis and 'rheumatic affections' due to their bathing in the spring waters 'as a pure matter of gratification, and not in any sense therapeutically' (AJHR, 1885: 10). Such engagements with these supposedly therapeutic blue spaces were deemed a breach of the correct self-care practices necessary to ensure good health, with emphasis placed on individuals being able to police their own behaviour and keep emotions (and impulses) in check. In the Rangitāiki wetlands, actions were taken from the late 19th century to restrict the ability of Māori to access and use geothermal resources as part of the colonial making at the same time Europeans sought to narrate and market therapeutic landscapes (such as spa towns) on the basis of past native 'health' histories, which paralleled processes which occurred to Māori and other indigenous peoples elsewhere (Foley et al., 2011: 153; Stokes, 2000).

In 1866, the settler colonial government, following on from the military invasion of the area in 1865 by colonial military forces assisted by Māori allies, enacted legislation that confiscated the majority of Rangitāiki wetlands from Māori iwi (Belich, 2015; Parsons and Nalau, 2016; Waitangi Tribunal, 1999). Confiscation (raupatu) was used by the colonial government throughout the North Island of Aotearoa New Zealand in the 1860s as a means to punish iwi who supposedly challenged British colonial rule, and also a mechanism to gain large tracts of Māori land for settlers (for further details, see Belich, 2001; O'Malley, 2016; Parsons and Nalau, 2016; Waitangi Tribunal, 1999). After the confiscation of Māori land and the imposition of colonial rule, the central government undertook land surveys in the Rangitāiki wetlands, and made leasehold and freehold lands available for European settlers in the 1890s.

## Government drainage in the Rangitāiki: 1910s–1940s

Between 1894 and 1910, European settlers attempted, unsuccessfully, to drain the Rangitāiki wetlands. In 1910, following repeated requests from local settlers for assistance, the central government introduced specific legislation (Rangitāiki Land Drainage Act, 1910) and provided funding (both directly through grants and indirectly through taxes on landholders) to facilitate large-scale engineering and drainage works in the wetlands (see Figure 13.2) (AJHR, 1911; Department of Lands and Survey, 1908; Law, 1962; New Zealand Parliament, 1910). As part of the wetlands drainage, river realignment and flood control operations, the lower portions of the Rangitāiki, Whakatāne and Tarawera rivers were almost entirely reengineered between 1910 and 1917 (AJHR, 1911, 1918) (see Figure 13.3a–c). Workers constructed canals using manual labour and new technologies

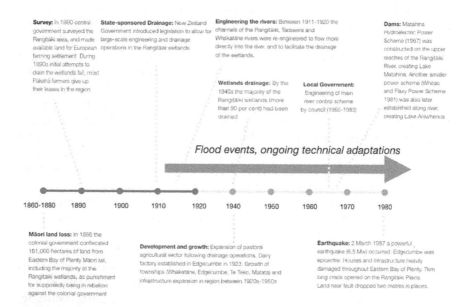

*Figure 13.2* Timeline of some of the changes to the Rangitāiki wetlands.
Source: Author.

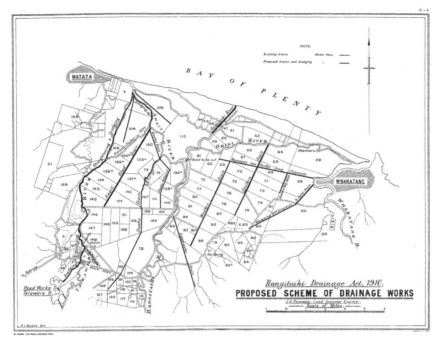

*Figure 13.3* (a–c) Government maps of Rangitāiki drainage scheme, 1911, 1917, and 1919.

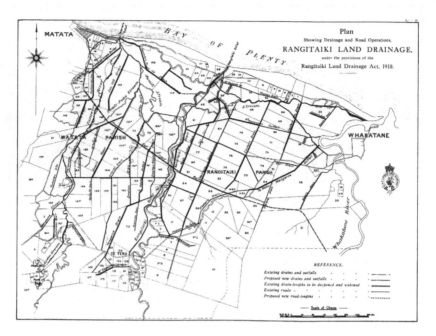

*Figure 13.3* (Continued).

(dredging machines specially imported from the US and England), straightened rivers, built levees and cut a new outlet for the Rangitāiki River through the sand dunes to the sea so it flowed more directly to the ocean (see Figure 13.4).

Amidst all the engineered structures and other fixtures of settler colonialism (fences, roads, cows and sheep, factories and townships), the issue of the persistent and resistant rivers and wetlands continued to be at the forefront of settler concerns. For many European settlers and visitors to the Rangitāiki Plains, the three rivers and wetlands were unlike their past experiences and personal expectations of rivers (AJHR, 1911; Department of Lands and Survey, 1908; Pontet and McCallion, 1964; Unknown Author, 1928). European rivers were recalled and imagined (and reimagined) as tamed, well-behaved and fixed healthy blue spaces, in contrast to the tortuous, unstable and (potentially) unhealthy blue spaces. In Britain, work to drain the East Anglia wetlands commenced in the 16th century and many settlers cited East Anglia as the model that should be adopted in Aotearoa New Zealand (Hursthouse, 1861; Park, 2002). The expansive, erratic and porous wetlands and rivers of Rangitāiki floodplains bore little likeness to the now common and ubiquitous engineered, regulated and straightened rivers of modern Europe. The construction of drainage canals, pumping stations, levees, factories, towns, roads and other infrastructure raised expectations among settlers that the rivers (once divorced from the muddy wetland interlopers) would be steady presences in the landscape. Such advantages in engineering technology and

*Figure 13.4* Dredging, Rangitaiki River, Bay of Plenty, 1910.
Source: Tonks, Hylton Gary, 1940–: Photographs of dredging of Rangitaiki – Whakatāne Rivers 1910; Ref: 1/4-016471-G. Alexander Turnbull Library, Wellington.

knowledge created confidence that many aspects of 'nature' (which was previously deemed capricious) were controllable and improvable for the advancement of settler society (Lavau, 2011; Webster and Mullins, 2003).

Drainage of wetlands and river 'improvement' works provided settlers with the possibility of correcting the unhealthy, undesirable and unpredictable tendencies of muddy blue spaces. Informed by British agricultural traditions, urban design and tenure systems and technologies, individual settlers and both local and central governments considered efforts to drain the Rangitāiki wetlands and control the three rivers to be of pivotal importance (Hinton, 1970; Opie, 1983; Parsons and Nalau, 2016; Webster and Mullins, 2003). Drainage operations were seen as vital necessities for the creation of temperate agriculture, the establishment of settlements and other developments. Most notably the creation and maintenance of healthy settler communities. Health and wellbeing, while not explicitly mentioned in legislation, underpinned policy discourse surrounding wetlands and rivers throughout the period. The Rangitāiki Land Drainage Act (1910) was reflective of central government's developing policy (Land Drainage Acts, 1893, 1904, 1908) towards Aotearoa New Zealand's wetlands in the first half of the 20th century (Park, 2001). There were four main tenets of this policy: first, the wetlands were 'useless' in their existing indigenous states and were only valuable for their potential as flat fertile land for farming; second, wetlands did not possess any scenic value and therefore were not part of emergent preservation campaigns (to conserve forests, birds and lakes for recreational purposes); third, that wetlands were legally potential land. This assumed potentiality meant that entitlements to wetlands (specifically Māori rights of access and usage) were deemed by the government and courts to transfer with titles to land (once 'unwatered'); and lastly, the development of wetlands into farmland was deemed of such national importance that it required central government intervention and funding to ensure its success. These tenets detrimentally impacted Māori communities in particular, who considered the (so-called unimproved) Rangitāiki ecosystems to be of immense value (economically, socially and culturally significant spaces in terms of spirituality, health and wellbeing).

In contrast to Pākehā, Māori did not perceive wetland drainage and water engineering works as 'improvements' on the existing waterscapes and landscapes. Throughout Aotearoa New Zealand, Māori regularly petitioned members of parliament and government officials about their loss of lands, resources and degradation associated with drainage and flood controls (Bamford and Brown, 1909; Hone Te Anga, 1914; Raukete te Hara, 1916). In these petitions, Māori challenged the ongoing privileging of scientific knowledge, technologies, values and authority over Māori knowledge, values, sociopolitical governance structures and ways of living (AJHR, 1923, 1927; Waitangi Tribunal, 1999, 2009). In addition to requests for the return of Māori lands and financial compensation for the loss of land and resources, petitioners expressed concern about their loss of access to resources as a consequence of government policies. Generations of Māori from the Rāngitāiki catchment wrote and spoke out about the negative consequences of drainage, flood controls and other hydrological interventions on their *rohe* (traditional lands and waters), including the decline in indigenous flora and fauna, and loss of sites of cultural significance. The decrease in tuna (freshwater eels), piharau (lamprey), upokororo (grayling), papanoko (torrentfish) and inanga (whitebait) as a consequence of hard adaptations, loss of habitat and pollution of waterways meant that Māori whanau, hapu and iwi were not able to undertake traditional harvesting practices (Downes, 1918; Park, 2001; Waitangi Tribunal, 1993, 1999). While central government officials did sometimes acknowledge that Māori petitions had merit, they repeatedly expressed the view that land development, drainage operations and flood controls were more important (economically, socially, politically and medically) to the nation and Whakatāne district than Māori livelihoods and wellbeing (Hone Te Anga, 1914).

European settler society's devaluation of wetlands resulted in the interconnected freshwater blue spaces not only being drained and re-engineered, but also being used as 'sinks' (disposal places) for waste products (Giblett, 2014; Pulido, 2017). Urban (town sewerage), agricultural (effluent from livestock and chemical run off from farms) and industrial wastes (from saw mills and dairy factories) were discharged into the rivers and dumped on top of the remaining wetlands (so-called 'land reclamation' works). Such actions inevitably contributed to declining ecological health, as well as human health, particularly among Māori communities for whom wetlands were mahinga kai (food-gathering sites). There were outbreaks of infectious diseases (including typhoid fever in 1914, 1919 and 1926) among Māori communities linked to the harvesting of shellfish contaminated by human effluent. In some instances, local government issued temporary bans on harvesting activities to prevent further infections and deaths (Ban on Shellfish, 1943; Death from Typhoid, 1926; New Zealand Herald, 1919; Poisoned Shellfish, 1943). Beattie and Morgan (2017) observed that colonial freshwater management schemes across the British Empire, incidentally or by design, impacted on people's resilience (their capacity to cope with and manage shocks and disruptions such as floods, infectious disease outbreaks or economic recession) (Beattie and Morgan, 2017; Berkes and Ross, 2013; Colding et al., 2003). Indeed, the unevenness of freshwater and land management schemes (which included wetland drainage and river 'improvements') meant that some social groups in colonial societies benefited whereas others did not. In Aotearoa New Zealand settler colonial policies (be it environmental, education, health or social welfare) more often than not privileged Europeans/Pākehā/whites and disadvantaged Māori and other non-whites (Came, 2014; Dow, 1999; McClure, 2013; Salmond, 2017).

## (Un)successful separation: maintaining the division between green and blue spaces, 1940s–2000s

By the 1940s, more than 90% of the Rangitāiki wetlands were drained. Government officials and local settlers (as Pākehā residents chose to self-identify up until the 1960s) declared the Rangitāiki drainage scheme a wholesale 'success' (AJHR, 1918: 192; Bay of Plenty Beacon, 1945a; Clarkson et al., 2013). One European visitor in 1928, for instance, recounted the 'transformation' of the Rangitāiki wetlands into grasslands as a story of a wasteland 'redeemed' through settler labour and scientific knowledge. The 'wilderness' of 'far-spreading area of marsh and lagoon and creek', 'threaded by muddy water courses' and 'eel-swamps' was, in less than two decades, 'unwatered with scientific skill, by canals and a network of deep drains … [and] straightened' river courses. The result was an 'expanse of rich grass land, with its grazing dairy herds, … orchards and homesteads' 'under cultivation and habitation, a region of industry and wealth, the home of scores of prosperous settler-families' (Unknown Author, 1928). This narrative of linear (settler colonial) progress mobilised individuals and institutions in a teleological expectation of irreversible social and ecological transformation, with advances in scientific knowledge and engineering technology providing the means to facilitate such radical changes to social-ecological systems (Veracini, 2010). Yet, this narrative of ongoing modernising progress (found on colonial productivist landscapes) did not allow for the possibility or acknowledgement of the (intended or unintended) negative consequences of actions to transform the existing hybrid muddy blue and green environments of the Rangitāiki and elsewhere in Aotearoa New Zealand.

From the perspective of Pākehā settlers, these new and improved blue spaces, however, remained problematic environments. Whereas in the

19th century the hazardous potentialities of wetlands centred on the health risks (associated with miasmas, dirt, germs) were the primary concerns of settlers and governments, by the early 20th century the risk of flooding (to property, lives, dry landscapes) was the principal concern. Settler place attachment and feelings of belonging were attached to the 'improved' landscapes that they, their families and neighbours had created, and thus the return of meandering, changeable, mobile blue spaces represented more than just an unruly annoyance (as was suggested by Lavau in regard to Australia's Goulburn River), but a physical and discursive threat the (imagined) settler colonial order of things (Lavau, 2011). Unruly blue spaces, in the form of raupo (indigenous reeds) growing on grasslands, and regular flooding events (which damaged roads, houses and infrastructure, and killed introduced grasses, crops and livestock) rendered the reconfigured Rangitāiki Plains (home to grids of canals, farmlands and physical structures) in a state of persistent unmaking, instability and unsettlement. On 24 February 1944, a regional newspaper reported, 'heavy rain from the high back country … transformed the three main rivers [of the Rangitāikii Plains] into roaring torrents of water which swept across the landscape in a destructive flood'. The rivers remained unruly, with the porous (and persistent) muddy blue spaces, threating the established and imagined Pākehā agricultural order of things, which drew strict lines between land and water. Farmers 'suffered serious losses, involving hundreds of head of valuable high producing dairy stock'. The most severe flooding occurred along the Rangitāiki River, where the river overflowed its levees and flooded the township of Edgecumbe (Bay of Plenty Beacon, 1945a). 'Settlers', the journalist reported, in Edgecumbe waged an 'all-night battle … against the flood water' using sandbags to fill the gaps in the levees. However, ultimately their efforts were futile and water 'inundated the[ir] …houses and shops' (Bay of Plenty Beacon, 1944a). In refusing to remain confined to the newly straightened and carefully engineered channels, the Whakatāne, Tarawera and Rangitāiki rivers flouted the newly founded boundaries (of fences, pastures, townships, roads and railways, land and waterways) and sought to reassert their muddy blue origins. The re-formed rivers and newly created grasslands, therefore, did not always accord to the 'settled expectations' of European settler communities of the Rangitāiki Plains (see Figure 13.5 showing the small community of Thornton on the banks of the straightened and leveed Rangitāiki River in 1965) (Lavau, 2011).

The mobility of the rivers and constant reoccurrence of muddy blue spaces (flooded pastures, regrowth of indigenous flora) caused European residents much distress (Bay of Plenty Beacon, 1944a, 1945a; Evening Post, 1925; Māori Television, 2017; Northern Advocate, 1925). Many farmers bemoaned the rivers as self-destructive perils that compromised productive agricultural land. In 1961, for instance, one 'wet country' farmer on the Rangtiāiki Plains described the need for 'constant vigilance by the occupier; if there is any easing up, the pasture will soon revert to rushes and weed grasses and

*Figure 13.5* Thornton, Bay of Plenty, road bridge over Rangitaiki River, looking inland in March 1965.

Source: Whites Aviation Ltd: photographs. WA-37436-F. Alexander Turnbull Library, Wellington.

the soil will become waterlogged' (Reynolds, 1961). The omnipresent threat of floods and the return of the former landscape (of wetlands, lagoons, meandering rivers and indigenous biota) served as further justification for decisions to favour engineered flood infrastructure over other approaches (such as spatial planning, different types of farming and housing designs). Indeed, the beginnings of path dependencies were evident in the design and implementation of the technocratic wetland drainage/river management scheme centred on engineering experiences. Such path dependencies (lock-in effects) were not only technological but also social. The ideology of improvement, and the emerging dependence of certain groups of people on such technologies fostered behaviours, practices and social expectations that in turn led to further demands for more engineering works (see Figures 13.2 and 13.6). European farmers in the Rangitāiki Plains, for instance, regularly petitioned officials and wrote letters to newspapers requesting the establishment and reinforcement of drainage and flood control works, which focused keeping unruly, fluid, muddy waters away from people, property and industry, rather keeping people away from rivers through spatial planning and other non-engineering measures (Bay of Plenty Beacon, 1945a, 1945b; Opus International Consultants, 1987; Pontet and McCallion, 1964). These actions to repeatedly intervene in the functioning of dynamic systems (climatic, freshwater, terrestrial, biophysical, ecological and social) frequently did not meet with people's expectations, and sometimes interventions resulted in the opposite outcomes, with negative consequences on communities.

*Figure 13.6* Photograph of a levee in Edgecumbe that rebuilt following the April 2017 flood.
Source: Author.

A wealth of geographical and environmental history scholarship demonstrates how the hydrological engineering water works that occurred within the Rangitāiki wetlands often served to heighten the vulnerability of communities to climate-related hazards (most notably floods and droughts) (Beattie and Morgan, 2017; Costanza et al., 2008; Lahiri-Dut, 2014; Laska and Morrow, 2006; Parsons and Nalau, 2016). Since wetlands traditionally absorb excess water and 'slow the speed and reduce the height and force of floodwaters' (Clarkson et al., 2013: 195), the loss of wetlands, combined with biota change, and the development of industrial structures and urban areas (in Whakatāne, Edgecumbe, Mātatā and Te Teko) on the floodplains, all contributed to significantly alter river flow and behaviour, and increased flood vulnerability. In the Rangitāiki Plains, there has been a pattern of persistent and worsening flood events coinciding with drainage works (see Figure 13.2); in April 2017, the Rangitāiki River breached one of its levees and flooded the township of Edgecumbe, which resulted in the evacuation of 1,600 people and widespread destruction of homes (see Figure 13.6) (Akuhata, 2017; Bay of Plenty Beacon, 1944a; Evening Post, 1925). Each flood event was met with political and public discussions about how to solve the flood hazard, government inquiries and ultimately the reinforcement of the existing institutional arrangements and engineering solutions (Bay of Plenty Beacon, 1945a; Department of Lands and Survey, 1925; Opus International Consultants, 1987; Rangitaiki River Scheme Review Panel, 2017). Indeed, as late as 2017, emphasis remained firmly fixed on structural interventions designed to separate land/water, despite Māori complaints that such flood infrastructures (rockwalls, levees) destroy tuna (eel) habitat and contribute to declining numbers of freshwater fishes (see Figure 13.6) (Askey, 2011; Bay of Plenty Regional Council, 2017; Te Pahiopoto Hapu, 2017).

Climate change necessitates that governments and communities contemplate alternative ways of managing not only existing flooding risk, but also projected changes in climate variability and environmental risks, and the consequences of environmental changes on people's relationships to places they value. This includes the impacts of sea level rise, the increased frequency and/or severity of extreme weather events (including drought, flooding and tropical cyclones) and increased incidence of water-/vector-/food-borne diseases due to warmer temperatures (Bennett et al., 2014; Curtis and Oven, 2012). The impacts of climate change may include the loss or radical changes to places (and biota) of economic, sociocultural and spiritual significance. As Māori experiences of environmental dispossession in the Rangitāiki wetlands attest, environmental changes inevitably result in unintended outcomes and can contribute to diminishing health and wellbeing when people's relationships to particular spaces of economic, social and spiritual significance are disrupted.

The recently established Rangitāiki River Forum, a new co-governance arrangement between various iwi and local governments that emerged following Treaty Settlements,[1] signals a potential shift away from engineering solutions, scientific knowledge and settler colonial values to encompass different knowledges, values and relationships in freshwater management (Rangitaiki River Forum, 2015, 2016). The Rangitāiki River Forum states in its guiding document that its purpose is to enhance and protect 'the environmental, cultural, and spiritual health and wellbeing of the Rangitaiki River and its resources for the benefit of present and future generations', no mention is made of wetlands (Rangitaiki River Forum, 2015). While water is a hot topic of discussion across academic, political, media and activist forums, murky water that is neither one environment nor another is not of particular interest. Wetlands remain largely at the periphery to discussions of healthy blue and green spaces in Aotearoa New Zealand. Yet these muddy blue spaces are highly contested and contestable spaces, with the majority of the country's townships and cities located on floodplains, and flooding the most common hazard. There is therefore a need to balance the diverse meanings and values attached to these spaces in ways that addresses the historical and contemporary environmental injustices experienced by Māori, as well as the current and future risks associated with climate change to people and other living beings (Bollen, 2015; Knight, 2016; Meredith, 2002; Salmond et al., 2014; Strang, 2014).

## Conclusion

In this chapter, I briefly discussed some of the multitude of hydrological interventions that were undertaken as part of the settler colonial project in Aotearoa New Zealand, but countless others (technical, ecological and sociopolitical) are outlined by other environmental historians and geographers including Beattie, Knight, Pawson, Holland and Roche (Beattie, 2003; Knight, 2016; Pawson and Holland, 2005; Peden and Holland, 2013; Roche, 1987). The colonial hydrological works undertaken in the Rangitāiki wetlands saw European settlers, engineers and governments imposing their scientific, technological and environmental knowledges, and sociocultural values onto the muddy blue spaces they encountered in the belief that they were spreading civilisation, progress, and correcting environmental deficits. Yet, those beliefs about wetlands and settler-led actions, however, resulted in the lasting negative outcomes (including the loss of resilience, increased flood risk, decreased biodiversity), which make the communities of the remade Rangitāiki Plains more vulnerable to the negative impacts of climate change.

This chapter is a small story of larger global stories about wetland loss and environmental changes. In writing it I sought to demonstrate the importance of telling 'small stories' about past experiences of environmental conditions and changes, which provide tangible connections to the past, and to complex mechanisms that might otherwise be beyond our ability to comprehend (Cameron, 2012; Morgan, 2013). By telling a story about how people understood, related to and interacted with muddy blue spaces in Aotearoa New Zealand, I wanted to explore the factors that contributed towards environmental loss and change (from the Rangitāiki wetlands to Plains), as well as to help make sense of what could be learnt to inform the ways in which we approach the uncertainties and anxieties associated with the Anthropocene.

## Note

1 A Treaty settlement refers to a legal package given by the New Zealand government to an individual Māori iwi, which includes a formal apology, financial compensation and other financial and legal components (including the right to buy back assets, the transfer of land and sometimes the right to co-govern parks, rivers or other sites of significance). Treaty settlements are a form of apology and reparations by the New Zealand government to a Māori iwi for historical injustices (including the confiscation of land, colonial violence, loss of resources, and racially discriminatory policies) committed by the government against Māori, which breached the articles of the Treaty of Waitangi. See Parsons and Nalau (2016).

## References

AJHR. (1885). *G-02a Reports from Native Medical Officers.* Wellington, Government Printer, AJHR No. G-02a.

AJHR. (1911). *C-11 Drainage Operations in the Rangitaiki Plains Report for the Year Ending 31 March 1911.* Wellington, Government Printer, AJHR No. C–11.

AJHR. (1918). *C-11 Rangitaiki Land Drainage: Report for the Year Ended 31 March 1918.* Wellington, Government Printer, Appendices to the House of Representatives No. C–11.

AJHR. (1923). *G-06f Native Land Amendment and Native Land Claims Adjustment Act, 1922. Report And Recommendation On Petition No. 187/1922, Relative To Successors Appointed To Interests of Rapata Nepia and Mereana Te Marohuia in Haupoto Block and Lots 28b and 31, Parish Of Rangitaiki*, Untitled, 1 January 1923. Wellington, Government Printer, Appendices to the House of Representatives No. G-06f.

AJHR. (1927). *I-03 Native Affairs Committee (Reports of the). Nga Ripoata a Te Komiti Mo Nga Mea Maori.* (Hon. Sir Apirana Ngata, Chairman), Untitled, 1 January 1927. Wellington, Government Printer, Appendices to the House of Representatives No. I-03.

Akuhata, K. (2017). Flood review – Need for Maori on panel. *Whakatane Beacon,* n.d.

Anonymous. (1879). Tourist notes on Hawke's bay. *New Zealand Country Journal,* 3 (2), 87.

Askey, Peter. (2011). *Edgecumbe Rangitaiki Plains Flood Mitigation Project.* n.p.

Ausseil, A.-G.E., Jamali, H., Clarkson, B.R. and Golubiewski, N.E. (2015). Soil carbon stocks in wetlands of New Zealand and impact of land conversion since European settlement. *Wetlands Ecology and Management,* 23 (5), 947–961.

Bamford and Brown. (1909). Bamford & Brown to the Kawa Drainage Board Clerk, 4 December 1909.

Ban on Shellfish. (1943). *Bay of Plenty Beacon,* 21 December, p. 4.

Bashford, A. (1998). *Purity and Pollution: Gender, Embodiment and Victorian Medicine.* London, Springer.

Bay of Plenty Beacon. (1944a). Rangitaiki Plains Flooded. *Bay of Plenty Beacon,* 25 February.

Bay of Plenty Beacon. (1944b). Rangitaiki Plains Flooded. *Bay of Plenty Beacon,* 25 February.

Bay of Plenty Beacon. (1945a). Future of Edgecumbe. *Bay of Plenty Beacon,* 4 March.

Bay of Plenty Beacon. (1945b). Rangitaiki Floods. *Bay of Plenty Beacon,* 13 November, p. 5.

Bay of Plenty Regional Council. (2017). *Summary of Decisions Requested By Persons Making Submissions on Proposed Change 3 (Rangitāiki River) to the Bay of Plenty Regional Policy Statement.*

Beattie, J. (2003). Environmental anxiety in New Zealand, 1840–1941: Climate change, soil erosion, sand drift, flooding and forest conservation. *Environment and History,* 9 (4), 379–392.

Beattie, J.J. (2005). *Environmental anxiety in New Zealand, 1850–1920: Settlers, Climate, Conservation, Health, Environment.* Dunedin, University of Otago.

Beattie, J. (2008). Colonial geographies of settlement: Vegetation, towns, disease and well-being in Aotearoa/New Zealand, 1830s–1930s. *Environment and History,* 14 (4), 583–610.

Beattie, J. and Morgan, R. (2017). Engineering Edens on this 'Rivered Earth'? A review article on water management and hydro-resilience in the British Empire, 1860–1940s. *Environment and History,* 23 (1), 39–63.

Belich, J. (2001). *Making Peoples: A History of the New Zealanders, From Polynesian Settlement to the End of the Nineteenth Century.* Honolulu, University of Hawaii Press.

Belich, J. (2015). *The New Zealand Wars and The Victorian Interpretation of Racial Conflict.* Auckland, Auckland University Press.

Bennett, H., Jones, R., Keating, G., Woodward, A., Hales, S. and Metcalfe, S. (2014). Health and equity impacts of climate change in Aotearoa-New Zealand, and health gains from climate action. *The New Zealand Medical Journal* (Online), 127 (1406), 16.

Berkes, F. and Ross, H., (2013). Community resilience: Toward an integrated approach. *Society & Natural Resources,* 26 (1), 5–20.

Bollen, C. (2015). Managing the adverse effects of intensive farming on waterways in New Zealand – Regional approaches to the management of non-point source pollution. *New Zealand Journal of Environmental Law,* 19, 207–239.

Bonnell, J.L. (2014). *Reclaiming the Don: An Environmental History of Toronto's Don River Valley.* Toronto, University of Toronto Press.

Came, H. (2014). Sites of institutional racism in public health policy making in New Zealand. *Social Science and Medicine,* 106, 214–220.

Cameron, E. (2012). New geographies of story and storytelling. *Progress in Human Geography,* 36 (5), 573–592.

Clarkson, B.R., Ausseil, A.-G.E. and Gerbeaux, P., (2013). *Wetland Ecosystem Services. Ecosystem Services in New Zealand: Conditions and Trends.* Lincoln, Manaaki Whenua Press, 192–202.

Colding, J., Elmqvist, T. and Olsson, P. (2003). Living with disturbance: Building resilience in social-ecological systems. In, F. Berkes, Colding, J. and Folke, C. (eds), *Navigating Social-Ecological Systems: Building Resilience for Complexity and Change.* Cambridge, Cambridge University Press, 163–185.

Costanza, R., Pérez-Maqueo, O., Martinez, M.L., Sutton, P., Anderson, S.J. and Mulder, K. (2008). The value of coastal wetlands for hurricane protection. *AMBIO: A Journal of the Human Environment,* 37 (4), 241–248.

Curtis, S.E. and Oven, K.J. (2012). Geographies of health and climate change. *Progress in Human Geography,* 36 (5), 654–666.

Death from Typhoid. (1926). *New Zealand Herald,* 30 July, 10.

Denyer, K. and Robertson, H. (2016). Wetlands of New Zealand. In, C.M. Finlayson, Milton, G.R., Prentice, R.C. and Davidson, N.C. (eds), *The Wetland Book.* Amsterdam, Springer, 1–15.

Department of Lands and Survey. (1908). Folder: Department of Lands and Survey Annual Report Swamp Drainage – Rangitaiki Plains 1908–1917. *R208194323, BAJ2, 15051, A1467, 8/d.* Archives New Zealand. Wellington.

Department of Lands and Survey. (1925). Land Drainage – Rangitaiki Flood Review 1925. *R22418099, ACGT, 18190, LS1, 563, 15/152.* Archives New Zealand. Wellington.

Dodd, M. (2010). *Effects of Industry on Maori Cultural Values: The Case of the Tarawera River.* Hamilton, Waikato University Creative Commons.

Dow, D.A. (1999). *Maori Health and Government Policy 1840–1940.* Wellington, Victoria University Press.

Downes, T.W. (1918). Notes on eels and eel-weirs (tuna and pa-tuna). *Transactions and Proceedings of the New Zealand Institute,* 50, 296–316.

Evening Post. (1925). Rangitaiki Settlers. *Evening Post,* 30 January, p. 7.

Flikke, R. (2016). South African eucalypts: Health, trees, and atmospheres in the colonial contact zone. *Geoforum,* 76, 20–27.

Foley, R., Wheeler, A. and Kearns, R. (2011). Selling the colonial spa town: The contested therapeutic landscapes of Lisdoonvarna and Te Aroha. *Irish Geography,* 44 (2–3), 151–172. doi:10.1080/00750778.2011.616059

Giblett, R. (2014). *Canadian Wetlands: Places and People.* Bristol, Intellect Ltd.

Hinton, C. (1970). *Engineers and Engineering.* Oxford, Oxford University Press.

Hone Te Anga. (1914). Hone Te Anga v Kawa Drainage Board – (1914) *33 NZLR 1139.*

Horrocks, M., Irwin, G., Jones, M. and Sutton, D. (2004). Starch grains and xylem cells of sweet potato (*Ipomoea batatas*) and bracken (*Pteridium esculentum*) in archaeological deposits from northern New Zealand. *Journal of Archaeological Science,* 31 (3), 251–258.

Hursthouse, C.F. (1861). *New Zealand: The 'Britain of the South',* London, E. Stanford.

Irwin, G. (2013). Wetland archaeology and the study of late Māori settlement patterns and social organisation in Northern New Zealand. *The Journal of the Polynesian Society,* 4 (122), 311–332.

Knight, C. (2016). *New Zealand's Rivers: An Environmental History.* Christchurch, Canterbury University Press.

Kolig, E. (2007). Freedom, identity construction and cultural closure: The *taniwha,* the *hijab* and the *wiener schnitzel* as boundary markers. In, E. Rata and Openshaw, R. (eds), *Public Policy and Ethnicity: The Politics of Ethnic Boundary-Making.* Basingstoke, Palgrave Macmillan, 25–39.

Lahiri-Dut, K. (2014). Beyond the water-land binary in geography: Water/lands of Bengal re-visioning hybridity. *ACME: An International Journal for Critical Geographies,* 13 (3), 505–529.

Laska, S. and Morrow, B.H. (2006). Social vulnerabilities and Hurricane Katrina: An unnatural disaster in New Orleans. *Marine Technology Society Journal,* 40 (4), 16–26.

Lavau, S. (2011). Curious indeed, or curious in deed? Some peculiarities of post-settlement relations with an antipodean river. *Australian Geographer,* 42 (3), 241–256.

Law, K. (1962). *Ruled By The Rivers: Tales of The Pioneer Days of Thornton and Rangitaiki District Compiled for the Thornton School's 50th Jubilee Committee,* Tauranga, The Bay of Plenty Times.

Māori Television. (2017). Edgecumbe residents angry at Council over flood damage [online]. *Māori Television.* Available at: https://www.maoritelevision.com/news/re-gional/edgecumbe-residents-angry-council-over-flood-damage, last accessed April 14, 2017.

McClure, M. (2013). *A Civilised Community: A History of Social Security in New Zealand 1898–1998.* Auckland, Auckland University Press.

Meredith, D. (2002). Hazards in the bog – Real and imagined. *Geographical Review,* 92 (3), 319–332.

Morgan, R.A. (2013). Histories for an uncertain future: Environmental history and climate change. *Australian Historical Studies,* 44 (3), 350–360.

New Zealand Herald. (1919). Typhoid among Maoris, New Zealand Herald, Volume LVI, Issue 17215, 17 July 1919. *New Zealand Herald,* 17 July.

New Zealand Parliament. (1875). Napier Swamp Nuisance Act 1875 (*39 Victoriae 1875 No 4*).

New Zealand Parliament. (1893) *Land Drainage Act 1893.*

New Zealand Parliament. (1904) *Land Drainage Act 1904.*

New Zealand Parliament. (1908) *Land Drainage Act 1908.*

New Zealand Parliament. (1910). *Rangitaiki Land Drainage Act.*

Northern Advocate. (1925). Rangitaiki Plains. *Northern Advocate,* 30 January, p. 5.

O'Malley, V. (2016). *The Great War for New Zealand: Waikato 1800–2000.* Wellington, Bridget Williams Books.

Opie, J. (1983). Environmental history: Pitfalls and opportunities. *Environmental History Review,* 7 (1), 8–16.

Opus International Consultants. (1987). Rangitaiki Plains: Effect on Flood Protection and Drainage – G. Pemberton, Engineering Manager, Bay of Plenty Catchment Commission. *R2020013, ABZK, 24420, W56448, 38.* 2 March, Archives New Zealand, Wellington.

Panelli, R. and Tipa, G. (2007). Placing well-being: A Maori case study of cultural and environmental specificity. *EcoHealth,* 4 (4), 445–460.

Park, G. (2001). *Effective exclusion. An Exploratory Overview of Crown Actions and Maori Responses Concerning the Indigenous Flora and Fauna, 1912–1983.* Wellington, Waitangi Tribunal.

Park, G. (2002). *Swamps Which Might Doubtless Easily Be Drained: Swamp Drainage and Its Impact on the Indigenous. Environmental Histories of New Zealand.* Melbourne, Oxford University Press, 176–185.

Parsons, M. and Nalau, J. (2016). Historical analogies as tools in understanding transformation. *Global Environmental Change,* 38, 82–96.

Parsons, M., Nalau, J. and Fisher, K. (2017). Alternative perspectives on sustainability: Indigenous knowledge and methodologies. *Challenges in Sustainability,* 5 (1), 7–14.

Pawson, E. and Holland, P. (2005). Lowland Canterbury landscapes in the making. *New Zealand Geographer,* 61 (2), 167–175.

Peden, R. and Holland, P. (2013). Settlers transforming the open country. In, E. Pawson and Brooking, T. (eds), *Making a New Land: Environmental Histories of New Zealand* (2nd ed.). Dunedin, Otago University Press, 89.

Poisoned Shellfish. (1943). *Bay of Plenty Beacon,* 21 December, p. 5.

Pontet, E.D. and McCallion, A.J. (1964). *Falling Leaves of Memory: As Gathered by the Eastern Bay of Plenty Federation of Women's Institutes.* Opotiki, Eastern Bay of Plenty Federation of Women's Institutes.

Pulido, L. (2017). Geographies of race and ethnicity II: Environmental racism, racial capitalism and state-sanctioned violence. *Progress in Human Geography,* 41 (4), 524–533.

Rangitaiki River Forum. (2015). *Te Ara Whanui o Rangitāiki – Pathways of the Rangitaiki.* Whakatane, Bay of Plenty Regional Council.

Rangitaiki River Forum. (2016). *Co-Governance Arrangements in the Bay of Plenty.* Whakatane, Bay of Plenty Regional Council.

Rangitaiki River Scheme Review Panel. (2017). Rangitaiki River Scheme Review – April 2017 Flood Event. Final Report as supplied to Bay of Plenty Regional Council. 18 September 2017. Bay of Plenty Regional Council, *Final Report Supplied to Bay of Plenty Regional Council.*

Raukete te Hara. (1916). Received: 4th September 1915- From: *Native Affairs Committee, House of Representatives.* Subject: Petition No. 237/15 Raukete te Hara and 27 others. For return of land taken into Rangitaiki Drainage Area.

Reynolds, W.M. (1961). Pasture establishment on swamp on a Bay of Plenty Dairy Farm. *Proceedings of the New Zealand Grasslands Association,* 23, 65–68.

Roche, M.M. (1987). *Forest Policy in New Zealand: An Historical Geography, 1840–1919.* Auckland, Dunmore Press.

Rockel, I. (1986). *Taking the Waters: Early Spas in New Zealand.* Wellington, Government Printing Office.

Romero Lankao, P. (2010). Water in Mexico City: What will climate change bring to its history of water-related hazards and vulnerabilities? *Environment and Urbanization,* 22 (1), 157–178.

Salmond, A. (2017). *Tears of Rangi: Experiments across Worlds.* Auckland, Auckland University Press.

Salmond, A., Tadaki, M. and Gregory, T. (2014). Enacting new freshwater geographies: Te Awaroa and the transformative imagination: Enacting new freshwater geographies. *New Zealand Geographer,* 70 (1), 47–55.

Stokes, E. (2000). *The Legacy of Ngatoroirangi: Maori Customary Use of Geothermal Resources.* Department of Geography, University of Waikato. Retrieved from https://researchcommons.waikato.ac.nz/handle/10289/6323

Strang, V. (2014). The Taniwha and the Crown: Defending water rights in Aotearoa/ New Zealand: Defending water rights in Aotearoa/New Zealand. *Wiley Interdisciplinary Reviews: Water,* 1 (1), 121–131.

Te Aho, L. (2014). 14 Indigenous laws and aspirations for a sustainable world. The Earth Charter. In, L. Westra and Vilela M. (eds), *Ecological Integrity and Social Movements.* London, Routledge, 169–180.

Te Pahiopoto Hapu. (2017). Te Pahipoto Hapu of Ngāti Awa Submissions to the Proposed Change 3 (Rangitaiki River) to the Regional Policy Statement, Submitted for publication.

Unknown Author. (1928). The Whakatane Valley and Rangitaiki Plain. *The New Zealand Railways Magazine,* 3 (8). Available online from: http://nzetc.victoria. ac.nz/tm/scholarly/tei-Gov03_08Rail-tl-body-d17-d7.html

Veracini, L. (2010). *Settler Colonialism: A Theoretical Overview.* New York, Springer.

Waitangi Tribunal. (1993). *Te Ika Whenua – Energy Assets Report.* Wellington, GP Publications.

Waitangi Tribunal. (1998). *Te Ika Whenua Rivers Report.* Wellington, Brooker and Friend Ltd.

Waitangi Tribunal. (1999). *The Ngati Awa Raupatu Report.* Wellington, GP Publications.

Waitangi Tribunal. (2009). *Urewera report–Part I.* Wellington, GP Publications.

Webster, B. and Mullins, S. (2003). Nature, progress and the 'disorderly' Fitzroy: The Vain Quest for Queensland's' noblest navigable river', 1865–1965. *Environment and History,* 9 (3), 275–299.

# 14 Conclusion

## New directions

*Ronan Foley, Robin Kearns,*
*Thomas Kistemann and Ben Wheeler*

## Introduction

In this short conclusion, we draw together some common ideas from the preceding sections and tease out recurrent themes that collectively set an agenda for ongoing work on water, health and place. The individual chapters very effectively illustrate why interdisciplinary approaches, experiential accounts and an ongoing concern with equity and justice matter for health in the context of watery places and spaces. In framing this book more broadly as an exploration of hydrophilia, we will briefly revisit those themes and additionally consider critical questions of data and method, theory and hybrid health and wellbeing outcomes that we feel can better uncover the character of healthy blue-space connections. All of us live near water, in Dublin, Auckland, Bonn and Cornwall, respectively. These locations lead us to understand the differences between inland/coastal, cold/warm, empty/crowded and urban/rural waters and how each shapes the phenomenological, affective and material elements of those connections. There is an emergent sensibility, linked to wider critical global concerns, that requires us to also consider some of the assumptions, even privileges that come with those connections, as well as a potential introspection associated with green and blue space research. Such critical concerns with aspects of neoliberalism, inequality, biopolitics, resilience and care inform new writing by critical health geographers (Brown et al., 2009, 2017; Crooks et al., 2018), as well as from wider cross-disciplinary collections within which medical humanities, creativity and hydro-citizenship play significant roles (Atkinson and Hunt, 2019; Brown and Peters, 2018; Finlay, et. al., 2015; Roberts and Phillips, 2018). All this new research explores interconnectedness and blue ecologies that emerge in complex and often contradictory ways from individual, communal and global lives and settings.

In considering the multidisciplinary perspectives that appear throughout the book, we are informed by writers from anthropology, leisure studies, sociology, environmental sciences, cultural studies, medical humanities, the arts and psychology as well as different strands of medical and health geographies. Each of the authors has, in their own chapters, collectively considered, either explicitly or implicitly, the idea of healthy geographies of blue space and this has been highlighted in the latter part of Chapter 1. Health geography has much to offer to support and enable the interdisciplinary integration of the many world views of blue space and human health, and in explicitly expanding from what is often a local, regional or national scope to one that is global. Evident in many of the chapters has been an explicit engagement with policy and the wider world outside of academia through studies carried out in collaboration with governments, agencies and authorities, and from a range of sectors across society. A critical, relational approach to hydrophilia, one that values individual and community experiences, perceptions and circumstances in impacting upon the health impacts of watery places, is clearly needed, alongside more typical biomedical and epidemiological paradigms. Work is underway attempting to develop more holistic strategies; for example, the Seas, Oceans and Public Health in Europe (SOPHIE) project aims to do just this in defining a research agenda for oceans and human health in Europe (https://sophie2020.eu/). Similarly, the Sustainable Development Goals (SDGs), taken collectively, provide an opportunity to consider the global interconnections between water and health through promoting actions that are mutually beneficial for human health and aquatic environments.

In developing these interconnections, we only scratch the permeable skin of hydrophilic experience to consider its depths and flows and the many different bodies and spaces that articulate a positive water-health relationship. A concern with depth is important in developing what might be termed 'bathymetric thinking', a three-dimensional approach that shifts across a hydrological spectrum from shallows to deep water and mimics a land-based deep mapping that combines humans with more-than-human water (Foley, 2017; see Chapter 5). Boundaries are rarely clearly demarcated and are dependent on mobile hydrographic aspects of climate, tide and catchment that blur lines across a profoundly relational space. Flows within blue space are also reflected in very different processes and wellbeing outcomes/experiences emergent within those shifting variable depths. This is evident not just in bodies and their relations with islands, beaches, waves and canals, as documented in Part II of the book, but also in the more cultural accounts linked to population and environmental health work found elsewhere in the text. How we experience nature, and how that in turn shapes our health, is of growing interest in public health. However, the as-yet relatively shallow emergence of blue space within public policy needs to deepen. New research shows how people experience different dimensions of water, in quite different health-enabling and health-disabling ways, either as icy blue (Findlay, 2018), outdoors/indoors in public pools (Ward, 2017; Watson, 2017) or in comparative evaluations of different blue spaces within cities (Völker and Kistemann, 2015). Much of the research on the health-promoting qualities of our blue environments fails to recognise the potential risks, whether to our own health and wellbeing or to the aquatic environments themselves. As noted in previous chapters, many inland blue spaces are associated with perceptions or narrative histories of hydrophobic outcomes, especially in relation to quarries, rivers and reservoirs. While not exclusive to inland waters, there are strong associations with negative physical (drowning, waterborne diseases) and mental (self-harm and suicide) health outcomes also emergent in the literature, and these are critical perspectives that also need development (Collins and Kearns, 2007).

While much of the (more traditional) work on human health and water is focused on risk (whether water insecurity, pathogen transmission or physical hazard), it does so sometimes without recognising the potential for health improvement given the right social, political and environmental circumstances and intervention. The focus in Part III on equity and spatial justice extends into more embodied/emotional accounts from voices of difference, around gender, race, disability and sexuality, with chapters specifically outlining Māori, Aboriginal, African, African-American and Black, Asian and Minority Ethnic (BAME) perspectives, though recognising that there are many other missing perspectives as well (Wiltse, 2010). New research from gendered blue space is evident in chapters in the book on swimming, surfing and access, yet there is considerable potential in detailed studies that explore gendered approach to water and how gender politics have been enacted in the production of blue space across different cultures and environments. From a disabilities perspective, it is also important to develop new and inclusive research, of which the Sensing Nature project at the European Centre for Environment and Human Health (ECEHH) is an excellent example with its focus specifically on how people with a range of sight-impairments use – or are enabled to use – blue and green spaces for healthy purposes. A fuller engagement with questions of equity and justice requires us to pay attention to differential capacities, individual and social, that construct water as an 'emotional and embodied equalizer', both in the sense of balance and fairness and as a potential protector for all.

## Critical questions and directions

The question of value represents a significant strand in current policy thinking on how natural spaces can protect and promote human health and this may inform data and methodologies associated with future hydrophilic research. The terminology of much of this debate draws upon environmental economics, a preference for quantifiable data and evidence (van den Bosch and Sang, 2017) and the use of the terms (cultural) ecosystem services and natural capital (Hegetschweiler et al., 2017). Such approaches have genuine potential in drawing attention to the value of such spaces, are taken relatively seriously and have the ear of public policymakers. Yet we argue that there are important elements of health-enabling blue space that are potentially excluded if the scope is confined solely to such instrumental and commodifiable approaches. A recent working paper identified that in London, city parks were valued on the asset registers as being worth a nominal £1, and there is an ongoing debate around how more realistic and meaningful valuations might apply to publicly owned blue assets (Orr et al., 2014). This is a genuinely thorny discussion that we feel future research should extend and develop, specifically in relation to data and methods that give a voice to qualitative and emotional responses to the meaning and value of places. Yet there is a need to augment and develop quantitative approaches rather than compete with them. Within this book, a quick scan reveals that the word 'value' appears around seven times per chapter. The term is used in different ways in relation to personal, social, economic and cultural values that suggest this is a complex and multifaceted issue and needs to be considered as such. Research that teases out what 'intangible' value really means to policymakers and develops alternative 'valuations' using innovative data and methods is badly needed. We suggest that combined work around psychological instruments, physiological markers and geo-narratives offers some ways forward, but additional in situ work and emplaced citizen science may tell us more about exactly how people evaluate blue space as meaningful to their and other's lives (Bell et al., 2015; Gidlow et al., 2016). But equally, and reflecting Anderson and Smith's (2001) succinct justification on why emotions matter in public policy, multiple narrative accounts of how different users (of all ages and types) use different blue spaces (of all shapes and sizes) will also be important. The ongoing Blue Health survey (https://bluehealth2020.eu/projects/bluehealth-survey/) is one example of an international comparative project seeking to do just that, and this kind of survey, augmented by local level qualitative research, has much promise.

While subtly explored in the book, we feel that many of the chapters identify and propose valuable theoretical lines of enquiry, supporting Andrews' (recent call for a fuller theoretical turn in health geographies (Andrews, 2018). Thinking around terms, such as flow, transmission, thalassography, minding, discovery, unpredictable/unpromising shades, authenticity or unwatering, opens up considerable scope for theoretical development not just in geography but beyond. Across the book, visibly relational elements are very well represented, as are different dimensions of a more-than-representational under-layer (Andrews et.al., 2014; Lorimer, 2005). Equally there are many examples of contested, even anti-therapeutic practices and experiences that emphasise blue space as an active subject component within a range of intersections from the disembodied and imaginative to the embodied and multi-sensual. Aspects of health psychology such as attention-restoration, immersive effects and nature-connectedness also emerge in different forms that open up a consideration of blue spaces as networked therapeutic assemblages (Duff, 2014; Foley, 2011). Equally, the affective side – always hard to measure and describe – is an essential aspect of evident phenomenological connections. Jennifer Lea's description of affect as 'triggerable capacities' emerges strongly from encounters in blue space of all types – seaside, coast, shore, lake, canal, river, stream. Linked to other theoretical work around atmosphere, accretion and place memory (Foley, 2017; Lea, 2018; McCormack, 2018), the development of such affective accounts reverberates across the book and emerges as mostly positive in their relations to health and wellbeing.

It is clear from the preceding chapters that human relations with blue spaces (and water more generally) are deeply anchored, and that while these spaces are often contested and can present hazards to health, they also present, in their hybridity, huge opportunity for salutogenesis. It is also clear that there is huge global socio-spatial inequity in the distribution of 'healthy waters' and opportunities to enjoy them, between countries and among population groups within them. There is therefore a need for research, scholarship and societal action that acknowledges the nuance of our relationships with waters that may be both 'healthy' and 'unhealthy'. Similarly, many of the chapters speak to the idea of public spaces and a hydrophilia enacted by multiple individuals/publics. But this is to ignore the appeal and stamp of the private blue and this should be explored further in critical ways around the resource impacts of private swimming pools (and closures of public ones), exclusivity in terms of resorts and the increasing privatization of the beach and coast (Kearns and Collins, 2016) or considerations of closed versus open blue spaces. In exploring these patchworks of public/private blue space, we may gain further insights into that fundamental question of human geography, why such spaces might matter in their potential to embed existing structural inequalities? In extending this question to include biopower there is an interesting irony around relationships between obesity, physical activity risk and public blue space. On the one hand, a focus on the positive benefits of embodied, emotional and experiential engagements within blue space is a strong message across the book. Yet equally they might be seen as part of wider 'individualisation/personalisation' agendas that shift public health policy out of statutory oversight and responsibility and back on to the individual (Rawlins, 2008). Equally public blue (and green) spaces come under risk-averse health and safety strictures that also place responsibility back on the citizen while avoiding any statutory responsibilities. Yet across this book, what is very evident are ongoing accounts of water/health engagements that are shared/social/communal and resistive as well. It is always important to frame those engagements against structural factors of inclusion, inequalities and the deeper production of 'real' things that fundamentally shape one's capacities to access such health-enabling places and spaces (Bell et al., 2018).

But it is not our intent to be too cynical or suspicious, especially when it comes to the capacity of blue spaces to expose us to something better in and for ourselves and others. There is something too in the developmental and discovery aspects of health within blue space, as settings that open people up to learned transformations outside the everyday that in turn situates and re-orientates the body in non-everyday settings. Phoenix and Orr's (2012) work on pleasure provides a well-argued account of an under-estimated but important aspect of the motivations for using green and blue spaces for physical activity especially for older bodies. This contention applies across all age groups and we would argue that the 'philic' in hydrophilia is entangled in the exhilaration evoked by water in the world around us – expressed in the global and joyful trope of the jumping/diving child or the floating 'oldie' drifting along a stream, echoing earlier reference to a *'hedonic turn'* (Lea, 2018). The joy of the blue is captured in accretion and expertise and embodied flow; in the water you potentially stop thinking and just *be* in a material and affective blending within an emplaced lifeworld (Phoenix and Orr, 2012). Extending this view out into the spaces themselves, we would propose a more holistic perspective that draws from newer cultural and economic geographic work on non-human geographies, blue economies and the need for a sustainable vision that mirrors current terms like One Health in the idea of 'One Blue' (Nicholls, 2014). This co-care of human and more-than-human health as a form of eco-social salutogenesis can be captured in something like 'beach clean' events, and might even extend to more numinous/spiritual aspects that, while underplayed in this book, do inform wider writing on the blue (Bell et al., 2018; Brown and Peters, 2018; Shaw and Francis, 2008).

In closing, we note that all critical human geography books should consider the hard question, why does all this matter? Based on the shared insight of the co-authors of this text, we argue that it matters for the following reasons:

- To provide a wider understanding of the relational ways in which healthy blue spaces are produced and sustained.
- To hear a fuller set of voices that explain the complex and differentiated ways in which people encounter health in blue space in previously under-considered ways.
- To extend instrumental measurement to consider more fully around how people value such places and use them in enabling and valuable health practices. This extended insight and knowledge needs to be built in to more formal valuations that preserve, protect and promote such spaces.
- To critically consider complex ways in which blue spaces are made available to societies for the creation of an effective public health.

To us, it feels like there is a new or shifting scale emergent across all of chapters in this book that considers, yet also fills out those complex spaces in, between and around the individual and societal. We feel that there is a crucial emergent scaling of community in its many different forms; as culture group, communities of interest and communities of practice that offer the best potential for more fully emplaced, political, performative and inclusive acts of hydrophilic care into the future.

# References

Anderson, K. and Smith, S. (2001). Editorial: Emotional geographies. *Transactions of the Institute of British Geographers,* 26, 7–10.

Andrews, G. (2018). Health geographies I: The presence of hope. *Progress in Human Geography,* 42 (5), 789–798.

Andrews, G., Chen, S. and Myers, S. (2014). The 'taking place' of health and wellbeing: Towards non-representational theory. *Social Science and Medicine,* 108, 210–222.

Atkinson, S. and Hunt, R. (2019). *Geohumanities and Health.* London, Springer.

Bell, S., Foley, R., Houghton, F., Maddrell, A. and Williams, A. (2018). From therapeutic landscapes to healthy spaces, places, and practices: A scoping review. *Social Science & Medicine,* 196, 123–130.

Bell, S., Phoenix, C., Lovell, R. and Wheeler, B. (2015). Using GPS and geo-narratives: A methodological approach for understanding and situating everyday green space encounters. *Area,* 47 (1), 88–96.

Brown, M. and Peters, K. (2018). *Living with the Sea: Knowledge, Awareness and Action.* London, Routledge.

Brown, T., McLafferty, S. and Moon, G. (2009). *A Companion to Health and Medical Geography.* (Blackwells Companion to Human Geography Series). Chichester, Wiley-Blackwell.

Brown, T., Andrews, G., Cummins, S., Greenhough, B., Lewis, D. and Power, A. (2017). *Health Geographies: A Critical Introduction.* Chichester, Wiley-Blackwell.

Collins, D. and Kearns, R. (2007). Ambiguous landscapes: Sun, risk and recreation on New Zealand beaches. In, A. Williams (ed), *Therapeutic Landscapes.* Aldershot, Ashgate, 15–32.

Crooks, V., Andrews, G. and Pearce, J. (eds) (2018). *Routledge Handbook of Health Geography.* London, Routledge.

Duff, C. (2014). *Assemblages of Health. Deleuze's Empiricism and the Ethology of Life.* New York, Springer.

Finlay, J., Franke, T., McKay. H. and Sims-Gould, J. (2015). Therapeutic landscapes and wellbeing in later life: Impacts of blue and green spaces for older adults. *Health & Place,* 34, 97–106.

Finlay, J. (2018). 'Walk like a penguin': Older Minnesotans' experiences of (non) therapeutic white space. *Social Science and Medicine,* 198, 77–84.

Foley, R. (2011). Performing health in place: The holy well as a therapeutic assemblage. *Health and Place,* 17, 470–479.

Foley, R. (2017). Swimming as an accretive practice in healthy blue space. *Emotion, Space and Society,* 22, 43–51.

Gidlow, C., Jones, M., Hurst, G., Masterson, D., Clark-Carter, D., Tarvainen, M., Smith, G. and Nieuwenhuijsen, M. (2016). Where to put your best foot forward: Psycho-physiological responses to walking in natural and urban environments. *Journal of Environmental Psychology,* 45, 22–29.

Hegetschweiler, T., de Vries, S., Arnberger, A., Bell, S., Brennan, M., Siter, N., Olafsson, A., Voigt, A. and Hunziker, M. (2017). Linking demand and supply factors in identifying cultural ecosystem services of urban green infrastructures: A review of European studies. *Urban Forestry and Urban Greening,* 21, 48–59.

Kearns, R. and Collins, D. (2016). Aotearoa's archipelago: Re-imagining New Zealand's Island geographies. *New Zealand Geographer,* 72 (3), 165–168.

Lea, J. (2018). Non-representational theory and health geographies. In, V. Crooks, Andrews, G. and Pearce, J. (eds), *Routledge Handbook of Health Geography.* London, Routledge, 144–152.

Lorimer, H. (2005). Cultural geography: The busyness of being 'more-than-representational'. *Progress in Human Geography,* 29 (1), 83–94.

McCormack, D. (2018). *Atmospheric Things: On the Allure of Elemental Envelopment (Elements).* Durham, NC, Duke University Press.

Nicholls, W.J. (2014). *Blue Mind. The Surprising Science That Shows How Being Near, In, On, or Under Water Can Make You Happier, Healthier, More Connected, and Better at What You Do.* New York, Little, Brown and Company.

Orr, S., Paskins, J. and Chaytor, S. (2014). Valuing urban green space: Challenges and opportunities. *UCL Policy Briefing,* October 2014.

Phoenix, C. and Orr, N. (2014). Pleasure: A forgotten dimension of physical activity in older age. *Social Science and Medicine,* 115, 94–102.

Rawlins, E. (2008). Citizenship, health education and the UK obesity 'crisis'. *ACME,* 7, 135–151.

Roberts, L. and Phillips, K. (eds) (2018). *Water, Creativity and Meaning: Multidisciplinary Understandings of Human-Water Relationships* (Earthscan Studies in Water Resource Management). London, Earthscan from Routledge.

Shaw, S. and Francis, A. (2008). *Deep Blue: Critical Reflections on Nature, Religion and Water.* London, Equinox.

van den Bosch, M. and Ode Sang, Å. (2017). Urban natural environments as nature-based solutions for improved public health – A systematic review of reviews. *Environmental Research,* 158 (Supplement C), 373–384.

Völker, S. and Kistemann, T. (2015). Developing the urban blue: Comparative health responses to blue and green urban open spaces in Germany. *Health & Place,* 35, 196–205.

Ward, L. (2017). Swimming in a contained space: Understanding the experience of indoor lap swimmers. *Health & Place,* 46, 315–321. Online corrected proof.

Watson, S. (2017). Liquid passions: Bodies, publics and city waters. *Social & Cultural Geography.* doi:10.1080/14649365.2017.1404121.

Wiltse, J. (2010). *Contested Waters: A Social History of Swimming Pools in America.* Chapel-Hill, University of North Carolina Press.

# Index

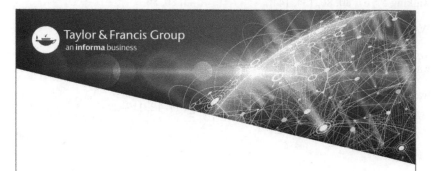

9 780367 661809